Wind From The East
Oriental Style Residence

深圳市创扬文化传播有限公司 编

赵欣 译

大连理工大学出版社
Dalian University of Technology Press

图书在版编目(CIP)数据

意蕴东方:东方风格样板间:汉英对照 / 深圳市
创扬文化传播有限公司编;赵欣译. — 大连:大连理工
大学出版社,2011.8

ISBN 978-7-5611-6372-6

Ⅰ.①意… Ⅱ.①深… ②赵… Ⅲ.①住宅—室内装
饰设计—中国—图集 Ⅳ.①TU241-64

中国版本图书馆CIP数据核字（2011）第149258号

出版发行:大连理工大学出版社
　　　　（地址:大连市软件园路80号　邮编:116023）
印　　刷:利丰雅高印刷（深圳）有限公司
幅面尺寸:242mm×280mm
印　　张:21.5
插　　页:4
出版时间:2011年8月第1版
印刷时间:2011年8月第1次印刷
责任编辑:刘　蓉
责任校对:李　雪
封面设计:连　帅
特约编辑:高　莹

ISBN 978-7-5611-6372-6
定　　价:298.00元

电　话:0411-84708842
传　真:0411-84701466
邮　购:0411-84703636
E-mail:designbooks_dutp@yahoo.cn
URL:http://www.dutp.cn

如有质量问题请联系出版中心:（0411）84709246　84709043

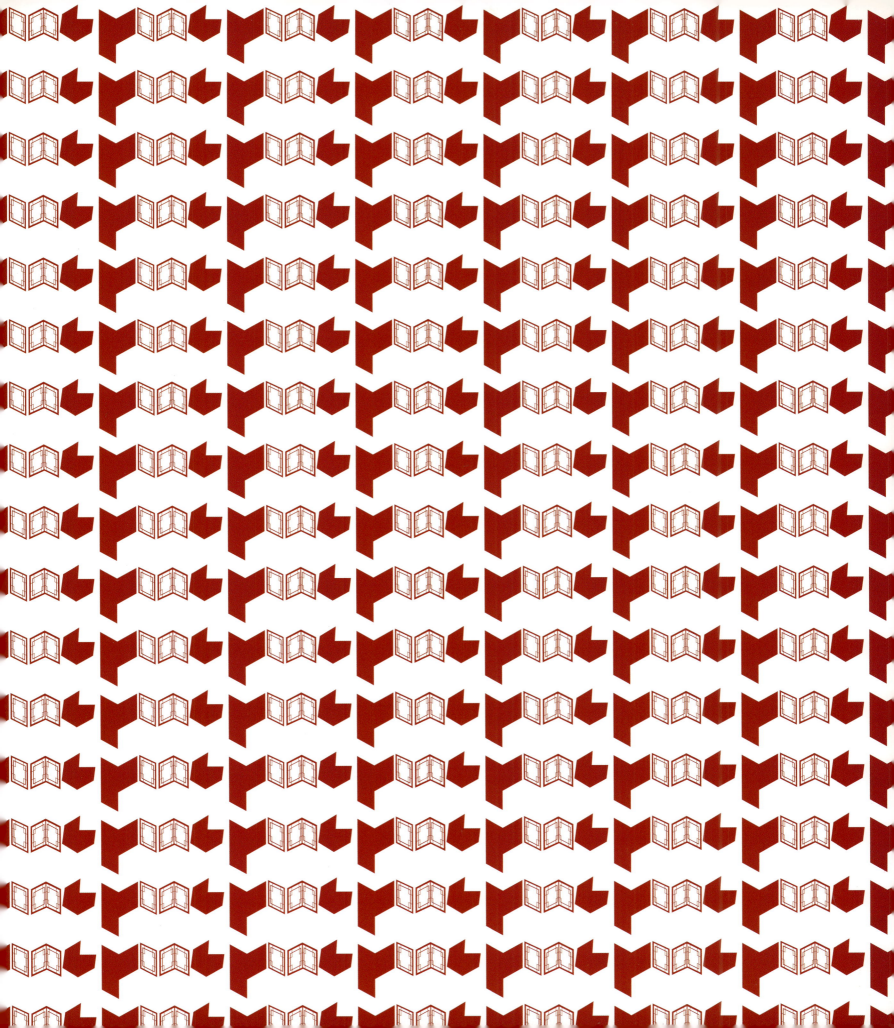

contents 目录

Charm · Jiangnan
韵 · 江南

This project is a single-floor residential space with a construction area of about three hundred square meters. The space is rather wide and the natural lighting condition is very good. For this reason, centering on its own advantages of this architecture, the designer develops a fashionable picture which takes the new oriental minimalism as the theme. In the process of his design, the designer has taken into full consideration of the balance relationship between the usage functions and visual effects, and manages to make the past serve the present. In addition to that, he combines the quintessence of the Chinese-style, traditionally classical architectural elements with the fashionable feeling of modern life into one. Apart from these, the designer also organizes the proportions, dimensions and material textures of the spatial configuration with the arrangement of colors in a harmonious way, so that it strives to achieve the reasonable innovation in the whole design, and also expects to represent the fashionable standards of the spirit of the time as well as the humanistic quality.

地点：福州　面积：300平方米　设计师：胡建国　设计公司：福州华悦空间艺术设计机构　主要材料：仿古砖、大理石、壁纸、玻璃、实木花格

　　本案为建筑面积约300平方米的单层住宅空间，空间较宽且采光条件良好，因此设计师围绕这一建筑自身的优势展开以新东方简约主义为主题的时尚画面，在设计过程中充分分析使用功能与视觉效果之间的平衡关系，并努力做到古为今用，将中式传统古典建筑元素的精华所在与现代生活的时尚气息融合在一起，并将空间造型的比例、尺度、材料质感与色彩搭配和谐地组织在一起，力求能在整个设计中达到合理创新，并能体现时代精神与人文品质的时尚标准。

Fu Tong Tian Yi Bay Show Flat, Dongguan
—Modern Chinese Style

东莞富通天邑湾样板房 —— 现代中式

Entering the door, your eyes will be caught by the tea pots of various forms that are placed on the wood grid partition. The Chinese-style furniture, the porcelain that can be seen everywhere, and the flowers that blossom on the screen ... the Chinese style atmosphere is embracing the whole space in a soft way. This project takes deep solid wood color and white color as the principal colors, and complements them with off-white color partially. Together with all types of Chinese style furnishings and ornaments, the designer creates a dignified, grand and smart space. There is no partition between the living room and the dining room, which chooses only wood flooring with varied colors as to divide the space, so that the visual space is thus expanded. The flowers that are placed on the dining table serve as an embellishment to the residential space, and the guqin placed on the balcony of the tea room brings a feeling of scholarliness to the space too. The walls of the master bedroom choose blossoming plum blossoms as the decoration, which just satisfies the psychological requirements of the occupant. While the bathroom employs marble and mirrors as the key decoration, which brings a somewhat different feeling to the whole space.

地点：东莞　面积：152平方米　设计师：韩松　设计公司：深圳市昊泽空间设计有限公司　主要材料：灰木纹石材、灰镜、紫檀木地板、紫罗兰木饰面板、布纹墙纸、手工刺绣

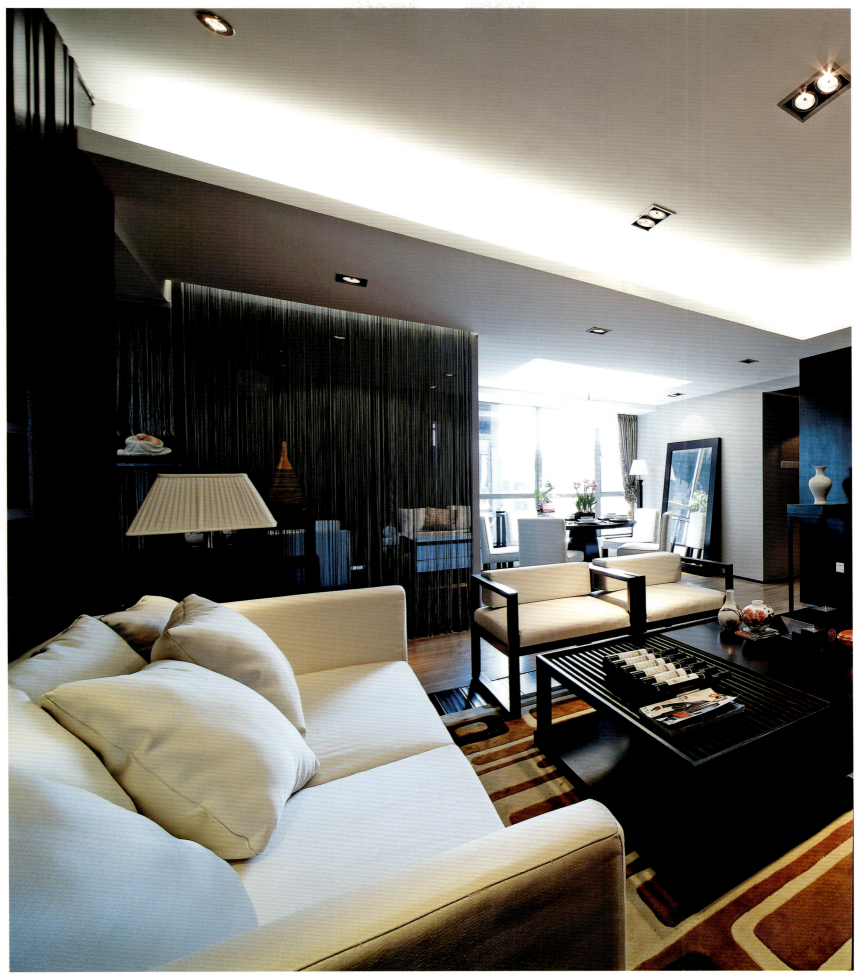

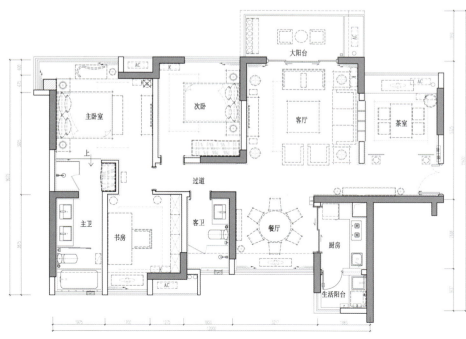

　　走进门，视线便被放置在木质方格隔断上造型各异的茶壶所吸引。中式的家具，随处可见的瓷器，绽放在屏风上的花朵……中式的氛围缓缓围绕着整个空间。本案以深实木色和白色为主色，用米黄色轻轻点缀，配合各式各样的中式饰品和摆设，打造出一个稳重端庄而又不失灵动的空间。客厅和餐厅之间没有设置隔断，而仅仅是以不同颜色的木地板作为空间的区分，拓展了视野空间。餐桌上摆放的鲜花，点缀了居家空间。茶室阳台上摆放的古琴，给空间注入了丝丝文化气息。主卧墙面用绽放的梅花作为装饰，满足了房屋主人的心理需求，而卫浴间以大理石和镜子为主，给整体空间带来些许不同的气息。

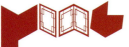

Fu Tong Tian Yi Bay Show Flat, Dongguan
— Southeast Asia No.1

东莞富通天邑湾样板房——东南亚1

This project is set as the Southeast Asian style, which takes warm colors as the key tone, and applies wood facing in a large area, such as the suspended ceiling and walls. Furniture also chooses the rich Southeast Asian-style Solid wood as the main material, such as the log-hued tea table and dining table. In terms of the soft decoration, the designer selects white color as the key tone, such as the cushions, curtains and bedspreads, ect, bringing a fresh air to the whole space, which is just in accordance with the designing theme of this project. The designer also craftily complements this project with some Southeast Asian style vinecirrus products, such as the couch cushions, the side tables and so on, so that the Southeast Asian national culture embraces the whole space in a skillful way.

地点: 东莞　面积: 89 平方米　设计师: 韩松　设计公司: 深圳市昊泽空间设计有限公司　主要材料: 浅橡木面板、席纹墙纸、米黄石材、灰镜、柚木地板

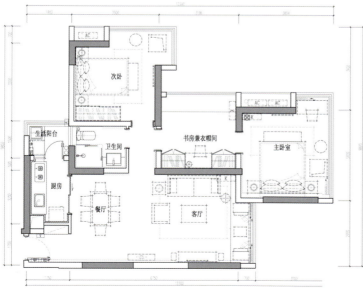

本案定位为东南亚风格，以暖色调为主，大面积的采用了木饰面，如吊顶、墙面。家具也以充满东南亚风情的实木为主，如原木色调的茶几、餐桌。在软装上，设计师以白色调为主，如靠垫、窗帘、床单等，给整个空间注入了一股清新之风，贴合了设计主旨。设计师还巧妙地加入了一些东南亚风情的藤蔓制品，如沙发靠背、边几等，使东南亚民族文化氛围巧妙地环绕了整个空间。

Fu Tong Tian Yi Bay Show Flat, Dongguan
— Southeast Asia No.2

东莞富通天邑湾样板房——东南亚2

This project is set as the Southeast Asian style, which in terms of design, gives much emphasis to the combination of the characteristics of Southeast Asian national islands and exquisite cultural taste. In this project, the designer applies a large amount of natural materials, such as wood, rattan and stone, and partially chooses ornamental decorations with brilliant colors. In the vestibule, the designer employs unsophisticated and spiffy layout as well as the spacious space which can provide eyes with a good ease and relaxation. The living room takes grandness and elegance as the principal tune. The sliding door with wood panes and the sofa background wood decorations are taken as the partition; as a consequence, cool and calm lines are used to partition the space, so that all over-complicatedness and obsessive decorations are thus taken place. The master bedroom selects log color as the key tune, and uses rattans to decorate the walls, so that the space is integrated but not monotonous. The bathroom space adopts the design of stone and mirrors, which enhances a lot of changes to the whole space.

地点：东莞　面积：108平方米　设计师：韩松　设计公司：深圳市昊泽空间设计有限公司　主要材料：深柚木面板、席纹墙纸、米黄石材、实木格栅、银箔、灰镜

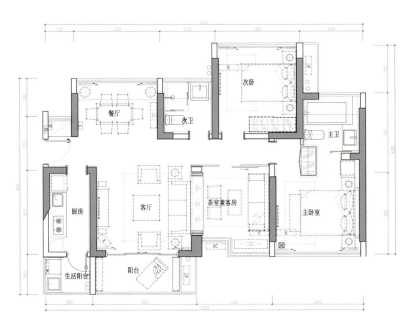

　　本案定位为东南亚风格，在设计上注重东南亚民族岛屿特色和精致文化品位的结合。本案广泛地运用木材、藤条和石材等天然原材料，局部用具有鲜艳色彩的饰品点缀。玄关处，简单利索的规划，宽阔的空间，使视觉得到舒展和放松。客厅以大气优雅为主，木质窗格的推拉门与沙发背景木装饰作为隔断，以冷静线条分割空间，代替一切繁杂与装饰。主卧室以原木色为主，局部以藤条装饰墙面，既统一又不会显得单调。卫浴空间采用了石材与镜子的设计，为整个空间增加了变化。

B2-7F Show Flat of Water Dynasty, Yuanli

元利水世纪B2-7F样板房

The living room of this project enjoys an excellent landscape advantage, due to this reason, the designer wants to lead such splendid view into the interior room. This project strives to place focus on the residence that possesses the leisure style, and puts much emphasis on the blending of Oriental cultural elements into the modern simplified style. As a consequence, many simplified and modern Chinese elements are craftily applied to the whole space without any trace.

The interior space plans out the living room, dining room, master bedroom, secondary bedroom, study room and public bathroom, as well as the kitchen. In such a limited space, it accurately works out the comfortable dimensions. The idea of planning the master bedroom derives from the layout of the luxurious rooms in hotels, which employs rather dark wallpaper to create a mind soothing atmosphere. What needs to be mentioned is the design of the study, which chooses cranberry glass to define the space, thus creating both the penetrability and the privacy. The corner near the window is designed to be a lounge, which can serve for multiple purposes, such as taking a rest, making a cup of tea and meeting guests, and it is an invaluable private space.

地点：中国台湾　面积：85平方米　设计师：谢启明　设计公司：大企国际空间设计有限公司

由于本案的客厅拥有极佳的景观优势，设计师希望将壮丽的景致引入室内，规划诉求着重于具备休闲风格的住居，并强调在现代简约的风格中融入东方元素，许多被简化、现代化的中国东方元素被巧妙而不着痕迹地运用在整体空间中。

室内空间规划了客厅、餐厅、主卧、次卧、书房及公共浴室和厨房。在有限的空间中，精密地计算出舒服的尺度，主卧房的规划构想源自饭店的奢华房型，采用较暗色的壁纸，营造出沉淀心灵的氛围。特别值得一提的是书房的设计，运用茶色玻璃作为空间的界定，兼具穿透性及私密性。靠窗的一隅则规划为卧榻，可同时兼具休憩、泡茶及客房等多重用途，是无价的私人空间。

Mu Lan, Taipei

台北沐兰

The designer craftily employs the remote and profound Oriental elements to encounter the modern and fashionable designing language. The brilliant colors mix-and-matched with materials of rich textures, fully develops the special sentiment of the Oriental culture. The designer places the visual focus into the space, such as the Chinese-style wood doors, lamps, furniture, birdcage and landscape, luxurious and chic. The nature of the space is elegant and tranquil, as the sober and reserved lines matched with the bright-colored blendent, and the soft touch of velvet mounting; the extreme tranquil elegance is interpreted in the current space and time, through the profound Oriental charm. The hazy Oriental and aesthetic space softly reveals the faint charm of different levels, so that when you are in it, you will feel like you are walking in a picturesque fairyland, heading into an enchanting illusion that wandering through the ancient and modern times.

地点：台北　设计师：杨焕生　设计公司：杨焕生建筑室内设计事务所

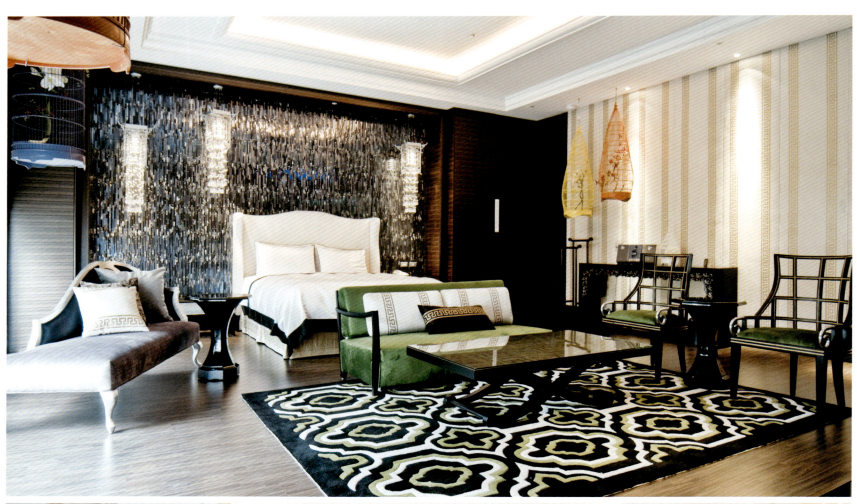

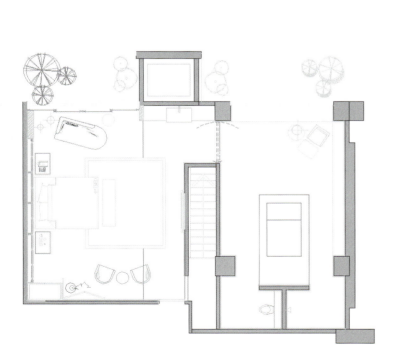

　　设计师巧妙运用幽远深长的东方元素，以邂逅现代时尚的设计语言，鲜艳的色彩与充满质感的才质精彩混搭，尽情演绎着东方的特有情味。设计师把中式木门、灯饰、家具、鸟笼、景观等视觉焦点缀饰于空间中，奢华而别致。空间本质是优雅静谧的，沉稳内敛的线条搭配高彩度配色，搭配丝绒裱布的细腻触感，将空间极度的清悠雅致，透过耐人寻味的东方意境，转译于时空当下。朦胧的东方美学空间，轻诉着幽微的层次韵味；置身其间，宛若画中情境，进入悠游古今的美丽错觉里。

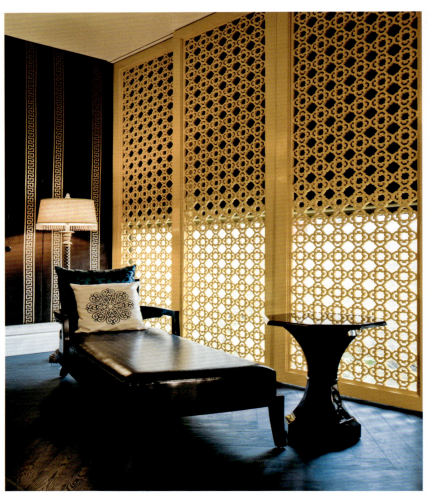

Type A of Jinsha West Garden Show Flat

金沙西园样板房A户型

The Oriental design style centers on the Oriental culture, and adopts modern designing approaches to embed modern designing ideas and inspirations in the design. The design not only inherits the Oriental culture, but also keeps pace with the times, combining the modern languages and modern techniques to rebuild the essence of the Oriental culture.

In terms of the materials, this project adopts a large quantity of wood, such as teak, deck plates and oak flooring, matched with grayish-green walls with plant fiber patterns, natural stone and wood carvings, to fully reveal the sense of mystery and naturalness of the Orientalism. At the same time, combined with modern materials, such as stainless steel and glass, it creates a tangible texture, and makes one feel more hospitable.

地点：成都　面积：130平方米　设计师：张晓莹　设计公司：多维设计事务所　主要材料：柚木、船甲板、橡木地板、有植物纤维纹理的墙纸、天然石材、不锈钢、玻璃　摄影师：赖珂

　　东方主义设计风格是以东方文化为核心，运用现代设计手法，将现代设计思想、设计感悟植入设计当中。它不仅是对东方文化的继承，同时与时俱进，集合现代语言和现代手法，重新构造东方文化的精髓。

　　本案在材料方面运用了大量木质材料，例如柚木、船甲板、橡木地板等，搭配带有植物纤维纹理的灰绿色墙纸和天然石材、木质雕花，尽显东方主义的神秘感与自然感；同时结合现代材质，例如，不锈钢、玻璃，形成可触摸的质感，令人倍感亲切。

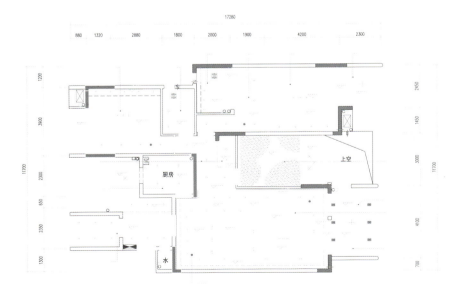

The Chinese-style Cheerfulness—Zhuo Yue Wei Gang Show Flat

中式的愉悦——卓越维港样板房

Life is not ready to be plain; therefore, the modern household life increasingly prefers to bring the past to the present, and applies modern techniques to endow a new aesthetic perspective, so that the tender cultural perception can be savored slowly. If a house is large enough, it fits in more of this performance of such feelings, which can more grandly blend all fancies for Chinese-style cultures, with a good structure, into the whole space plan and layout. In addition to that, it even chooses different materials to create a strong visual impact. Like in this project, though the foyer garden takes fashionable luxury as the design concept, the ceiling, floor and walls are all integrated into the inherent modern Chinese elements with hidden lines, which freely transfer and alternate in both the exterior and interior spaces. The Chinese-style environment also gives emphasis to the treatment of details, and takes the essence of classic furniture but creates with modern materials. In terms of colors, they can be brisk and bright colors, which compared to classic flavors are easy, cheerful and far-reaching in meaning. Like the tables and chairs which imitate the Qing Dynasty in the living room and dining room, the steel matches well with leather, and they look like antiques but more handy. Any functional space leaves only the comfortable touch that is mostly needed by bodies, such as the basin and bathtub, which will make one feel the safest and freest relaxation. From time to time, with some unique landscape ornaments, such as the large glass, it just enlarges the corner of space in an appropriate way.

地点：深圳　面积：180平方米　设计师：戴勇　设计公司：戴勇室内设计师事务所　主要材料：米色云石、黑金花云石、铁刀木、茶色玻璃、茶镜、艺术马赛克、壁纸

生活不会甘于平淡，现代家居生活，越来越喜欢引古于今，用现代手法赋予其全新的审美角度，慢慢品味其中细腻的文化感受。如果一个房子够大，就更适合这种情感的表现，可以更大气地将所有对中式文化的情有独钟层次分明地融入到整体空间规划与布局中，甚至利用不同材质去产生强烈的视觉碰撞，就像这间住房，入户花园虽以时尚奢华为设计理念，但天花、地面、墙体都以一条隐线统一于内在的现代中式元素，在室内外空间中自由转换和穿插。中式环境还强调细节的处理，取经典家居之神髓而以现代材质予以创作，色彩也可跳跃为鲜活的亮色，较之古韵觉着轻松欢愉且蕴意深浓，好比客厅、餐厅里那些仿自明代的桌椅，钢材与皮革配合，模样似曾古物，却更为轻巧。任何一个功能空间，只留下身体最需要的舒适接触，如洗手池、浴盆，让人享受最安全、最自在的放松。偶尔来点与众不同的景观点缀，如硕大玻璃，则恰到好处地放大了空间的角落。

Dark Green Fragrance

墨绿色的芳香

The designers choose dark green as the main color of the whole space, bringing distinctive flavor to the space. The purple yarn curtains wrap around the wooden beams, just like Thai silk with aristocratic atmosphere exuding its unique charm. A national embroidery with showy colors is placed randomly on the tea table as an alternative to local yarn curtain with Southeast Asian atmosphere, attached to the light teak lines of the tea table. The solid and rattan furniture are mixed and combined together, creating a whole Southeast Asian style and exuding the simplicity of earth, and the round and exquisite rattan furniture demonstrates the quality of home. Red cloth decorations with strong contrast to dark green are chosen to liven up the indoor atmosphere. The washroom is decorated with bright-colored mosaic with natural atmosphere to highlight the whole style, and the rough texture of the surface looks like everything after the rain, clear and bright, bringing deep tropical feeling.

地点：成都　面积：140平方米　设计师：廖志强、张静　设计公司：之境室内设计事务所　主要材料：乳胶漆、地砖、大理石、木材、壁纸

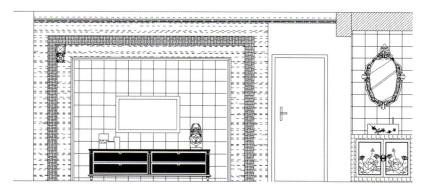

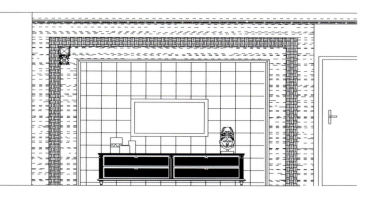

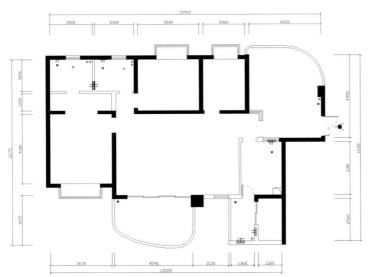

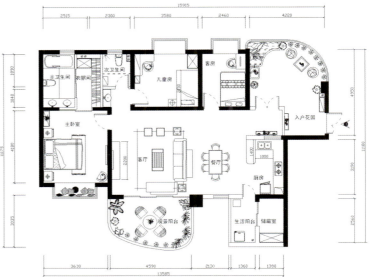

　　设计师选用墨绿色作为整个居室的主色，使室内风格别有风味。紫色纱幔缠绕在木梁之上，犹如带有贵族气息的泰丝散发出独特的韵味。茶几上随意摆放的一条色彩艳丽的民族绣品作为替代东南亚氛围的纱幔，附着茶几浅浅的柚木纹路。

　　实木和藤制家具的混搭组合，营造出整体的东南亚风格，散发着泥土的质朴气息。浑圆精巧的藤制家具更彰显了居家品质。选用和墨绿色有着强烈对比的红色布艺点缀，活跃了室内的气氛。卫生间使用了色彩亮丽、颇有大自然气息的马赛克来突出整体风情，表面粗糙的质感犹如雨后的万物，清晰明朗，带来浓浓的热带感受。

7#2-7-2 Type E Show Flat,
Jin Di Evergreen Bay, Shenyang

沈阳金地长青湾7#2-7-2 E户型样板间

The style of this show flat is set as the natural Orientalism, which takes the Southeast Asian style as the fundamental key to uniformly embody the essence of the Oriental . The designer uses spure, native and modern designing languages and elements of the Southeast Asia holiday paradise, such as the wood lattices, the ceiling beam, the flower-patterned suspended ceiling, and the Thai and Indonesian modern furniture and ornaments made of leather and rattan. When it comes to the creation of space, this project inherits the natural, healthy, and leisure qualities, from the creation of space, to the details decoration, it incarnates the respect and harmony to nature, and produces an Oriental poetic space and the rich and comfortable romantic appeal.

No matter in terms of the designing concept, or the sound application of materials, colors and the large quantity of plants, this project totally corresponds with the low-carbon life attitude that is attracted toward nature, and helps to realize northerners' longing and imagination for the southeastern living environment. As a consequence, the designer, in his designing language, expresses people's minds and meets the requirements of modern people.

The designer strives to create a comfortable and exotic mood,and this is prinapal aim of appeal of the project. As a result of this, the designer leaves romance everywhere, so that every scene is uniquely impressive; for this reason, the natural Oriental style of this project is perfectly presented ahead of you.

地点：沈阳　面积：约180平方米　设计师：陈贻、张睦晨　设计公司：北京睦晨风合艺术设计中心　　主要材料：地面：大理石、实木木地板　天花：立邦漆、泰柚木饰面、席编壁纸
墙面：麻编壁纸、米白洞机刨石、热带雨林啡石材、椰壳马赛克、艺术玻璃

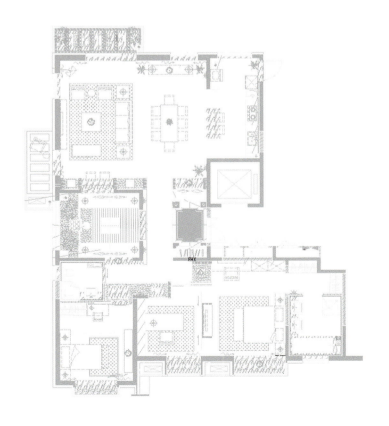

此套样板间风格定位为自然东方，以东南亚风格为基调，将东方人的精髓统一体现出来。设计师运用了纯粹、地道的并且是现代的东南亚度假胜地的设计语言和元素，如木质格栅、天花木梁、花格吊顶，以及泰国以及印尼的皮制和藤制现代家具和配饰。在空间营造上，本案继承了自然、健康和休闲的特质，大到空间打造，小到细节装饰，都体现了对自然的尊重与谐调，营造出了东方的诗意空间和浓郁舒适的浪漫情调。

本案无论是在设计理念方面，还是在材质、色调以及大量植物的合理运用方面，都完全切合了低碳、向往自然的生活态度，也成全了北方人对南方生活环境的向往和遐想，设计师用设计语言表达了人们的心思，符合现代人的需求。

设计师极力营造舒适度和异域情调，是设计的主要诉求目标，设计师处处留情，景景入心，将此样板间的自然东方风格完美地呈现在您的眼前。

The New Chinese Style of Zhu Guang New City Show Flat

珠光新城样板房之新中式

In this type of project, the designer sets its orientation as the new Chinese Style, which takes black, white and gray as the fundamental tone, creating a stable and elegant residential feeling. In the basement, the audio-visual room and the tea room have equal shares of the space. Considering the ceiling is rather low, the designer arranges a dry landscape between the audio-visual room and the tea room, which not only realizes the transition of the two spaces, but also gives consideration to the lighting requirements of the bathroom.

Ascending along the steps from the basement, you will come to the dining room, in which the application of Chinese-style zig-zaging copper coin patterns fills the whole background wall of the dining room, presenting the architectural symbols of the traditional Chinese culture. The white copper coin patterns make a strong contrast with the dark hickory wainscot board, which, along with scattered sparkling lights, seems to present a clips show of both the modern and old times. The hickory has plain wood textures that are similar to rosewood, so that it creates a traditional cultural atmosphere.

The living room takes advantage of the treatment of mirrors, so that the space is extended to a large degree. An outward double sliding door fairly separates the study room, in which the table and chairs are still Chinese style, and the customized Chinese-style bookcase, which is made on the spot, is arranged as the background of the study room at the back of the table.

面积: 244平方米　主案设计师: 霍承显　软装配饰师: 潘敏意　设计公司: 空间印象　主要材料: 银镜、茶镜、钢化玻璃、雅士白大理石、古木纹、马赛克、乳胶漆、木地板等

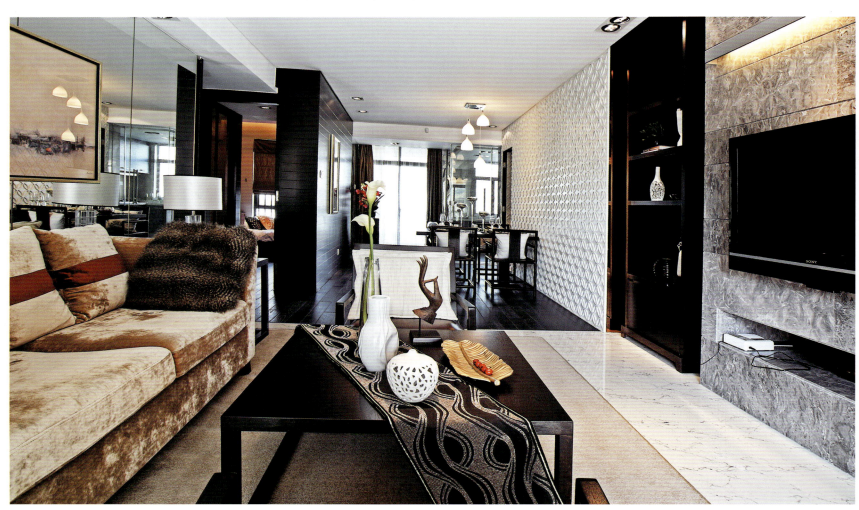

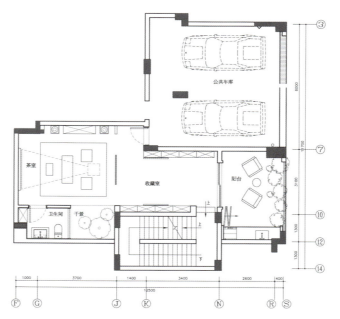

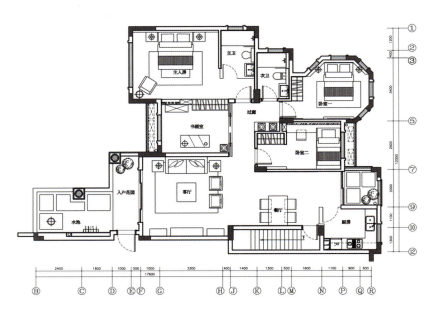

　　本套户型中，设计师以新中式的风格作为设计定位，以黑、白、灰为主色调，营造了一种平稳、雅致的居家感。地下室中影音室和茶室平分秋色，考虑到天花偏低，设计师在影音室和茶室之间做了一个内庭院的干景，既实现了两个空间的转换，又很好地兼顾了卫生间的采光需求。

　　沿着地下室拾级而上，就到了餐厅，中式的回形金钱纹运用浮雕手法铺满了整面餐厅的背景墙，显示出中国传统的建筑符号。白色金钱纹墙面与深色的桃木护墙板的反差对比，加上零星点缀的灯光，仿佛上演着现代与旧时岁月的片段剧。桃木那质朴的仿红木纹理营造出一种传统的文化氛围。

　　客厅通过镜面的处理使得空间大尺度地延伸。一扇外向双开推拉门将书房很好地区隔开来，依然是中式的书桌和椅子，量身定制设计并在现场打制的中式书柜，衬于书桌后面做书房的背景。

Refreshing Blue

沁兰

Once in the space, a rich spaciousness and warm atmosphere of home come to our faces, and the refreshing blue-green color is randomly dotted on the white background. The landscape painting, flower-bird painting, Buddha's head sculpture, and bird-cage-shaped flower shelf are hidden in the classic white harmoniously. The wallpaper with flower patterns and carpet with natural patterns spread to every corner of the bedroom, making the bedroom glow with a quiet but lush vitality. For the piece of pure white, the arhat bed, armchair, wood-carved flowers, cornice-shaped cabinet with seven drawers, dining table and chairs with flower and bird patterns and other furniture with traditional Chinese style show out an elegance beyond vulgarity and demonstrate a low-key luxury.

地点：武汉　面积：79平方米　设计师：周炀　设计公司：尚映空间　主要材料：爵士白大理石、密度板雕花、烤漆玻璃、银镜马赛克、艺术墙纸

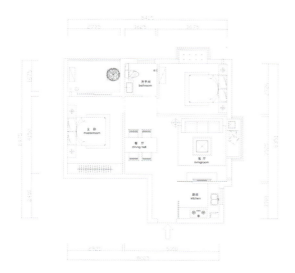

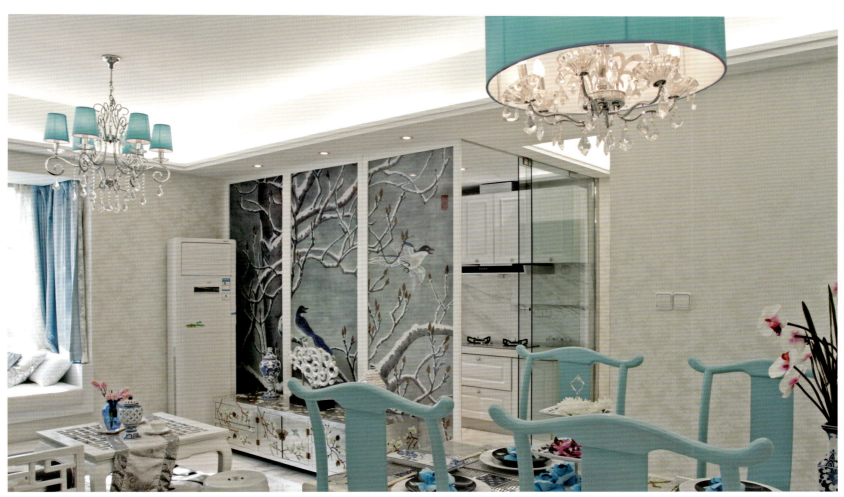

　　推开门，迎面扑来的是浓郁的空灵而温馨的家的气息。沁人心脾的蓝绿色随意地点缀在白色空灵的背景之中。山水画、花鸟画、佛头造型、鸟笼造型的花架，都和谐地隐匿在了这一片典雅的纯白之中。花草的壁纸、自然纹样的地毯蔓延至居室的每个角落，使居室焕发出宁静而茂盛的生命力。罗汉床、太师椅、木雕花、飞檐七斗柜、花鸟的餐桌椅等传统中式造型的家具，却因那一片纯白，有了超凡脱俗的典雅，彰显着低调的奢华。

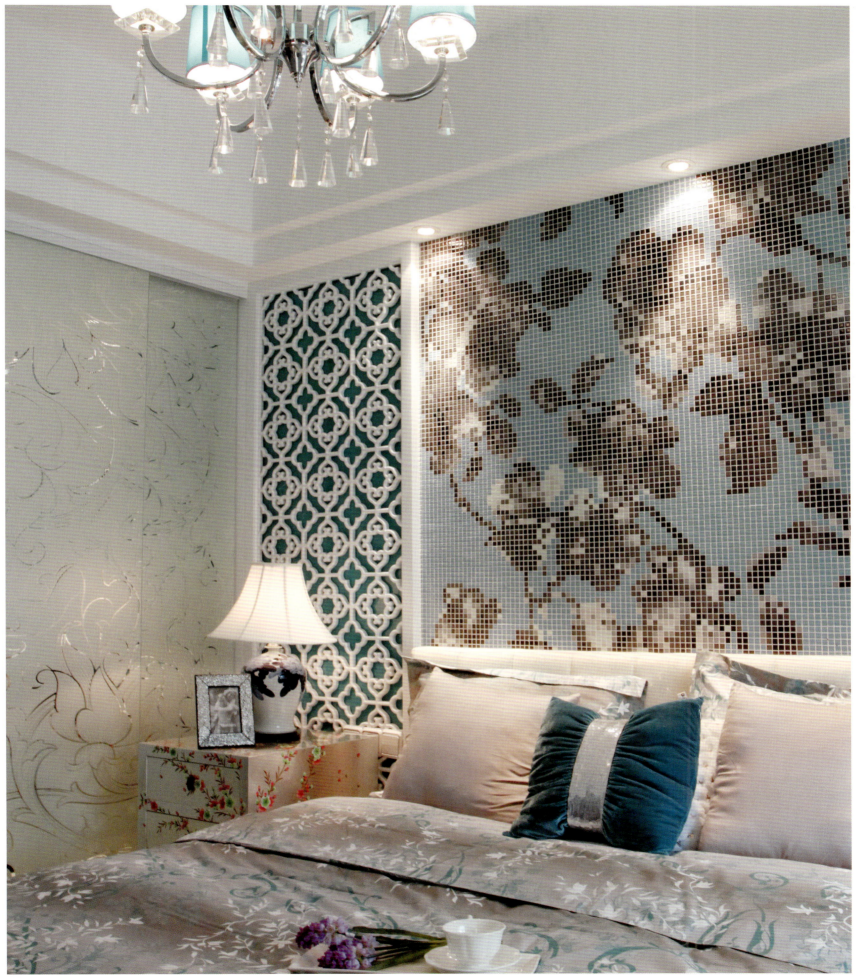

Relle
颐莲

This is a space full of taste of concept, so the designer named it "Relle", and the concept of Zen is introduced into the design of the space, embracing with the low-key details in the space. The pattern of the interior is square with open space, and the designer strives to show the simple mood in building the interior atmosphere. The dining room is connected with the living room, making the space opener, and one side of the TV wall is designed to be white with decorations of Chinese character component forms, enhancing the Zen concept of the space. The sofa background wall is hollowed with Chinese patterns, adding the traditional meaning to the space. One side of the walls in the dining room is decorated with a mural of lotus, and the ceiling in the living room is decorated with concave-convex lotus-leaf lines, closing to the design theme in all details.

地点：江苏常熟   面积：270平方米   设计师：官艺   设计公司：苏州绿松石室内设计工作室   主要材料：斑马木、白色人造石、热弯玻璃、马赛克、黑玻璃珠墙纸、灰镜

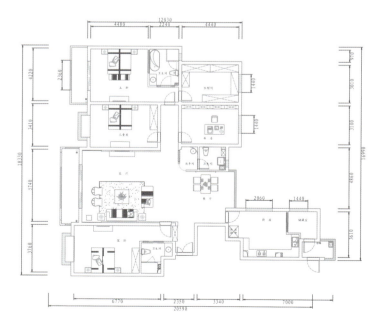

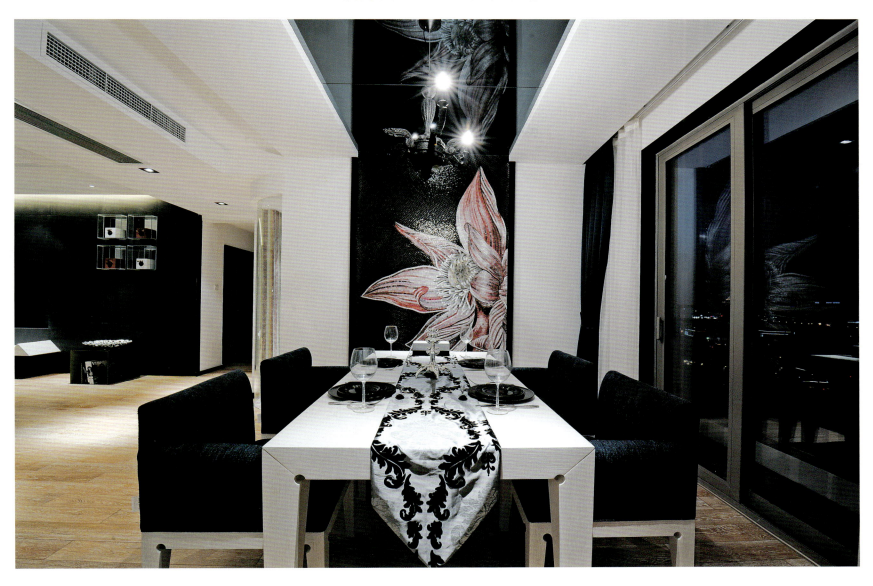

　　这是个充满概念味道的空间，设计师将其命名为"颐莲"，在空间的设计中融入禅的概念，空间中低调的细节相互辉映。室内的格局方正，空间开阔，设计师在室内氛围的营造方面，力求体现简约的意境。客厅和餐厅相连，使空间更为开阔。电视墙一侧的白色装饰墙面采用汉字偏旁形式的装饰，增添了空间的禅意。沙发背景墙的中式纹样镂空处理，增添了空间的传统意蕴。餐厅一侧的墙面是一幅莲花的壁画，客厅天花采用凹凸的荷叶线条，在种种细节中紧贴设计主题。

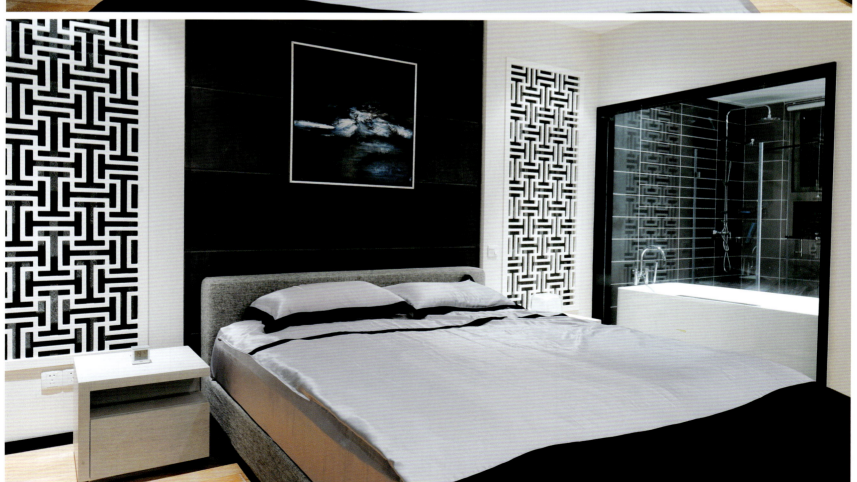

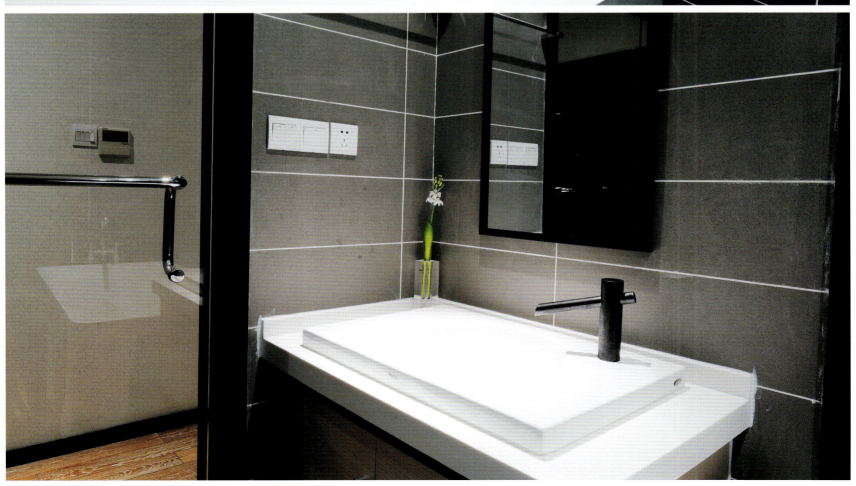

Oriental Garden, Xiangtan, Hunan

湖南湘潭东方名苑

Going into this new Chinese-style house, as if we go into a Chinese painting, with white rice paper as base below black ink, following the simple classical style and simplifying the complication.

The ground at the entrance is designed with simple brick collages, with a arhat-bed chair for changing shoes, both functional and ornamental. In the living room, the blue-flower skirting on the ground and the simple and delicate droplight set each other off on the white wall. The elegant ink flamed board is introduced as the partition between the living room and dining room, and we can see the new Chinese-style rosewood dining chairs through the grille of wooden pillars. In the choice of furniture, the designer deliberately avoids the treatment of rosewood with carvings all around, but chooses an increasingly tough handling way without losing the generous style.

地点：湘潭　面积：260平方米　设计师：张燕　设计公司：自在天装饰　主要材料：巴西花梨木、菠萝格、芝麻黑花岗岩火烧板、美生雅素丽砖、科勒洁具、格莱美墙纸、京瓷家具、创意布艺、松岚油画

　　走入这套新中式的房子，好像走入一幅中国画。白宣为底，黑墨其上，沿袭了古朴，也简化了繁琐。

　　入口处地面上做了简单的地砖拼贴，罗汉床形式的换鞋椅，兼顾功能性与观赏性。居室内地面青花色的角花与简洁精致的吊顶，共同映衬在墙面的白色间。客厅与餐厅的隔墙采用雅致的墨色火烧板，透过木柱的格栅可以见到餐厅花梨木的新中式餐椅。在家具的选择上，刻意避免了花梨木四处雕花的处理，反而采用更加硬朗的处理手法，却没有缺失丝毫的大气风范。

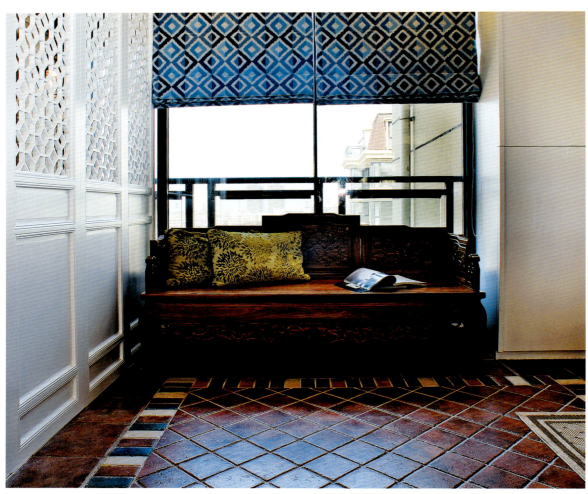

The Nostalgia for the Old House in an Apartment Building

公寓楼里的老宅情怀

Turning an armored concrete building into an old curtilage that is filled with archaic rhyme is the original designing intention of the designer. The designer particularly applies the construction elements of the Chinese-style architecture as the principal designing method, in the hope that the project will not leave any trace, nor be over-decorated, thus to create the Chinese-style charm. By using the eight-angled door, the fan-shaped windows, and miss windows, the designer has realized the view borrowing for every space, and achieved the effect of leading to a secluded quiet place through the winding path.

地点：福州　面积：280平方米　设计师：卓新谛　设计公司：福州合诚美佳装饰公司　主要材料：青砖、老榆木楼梯及栏杆、复古砖、草编墙纸、水曲柳面板　摄影师：施凯

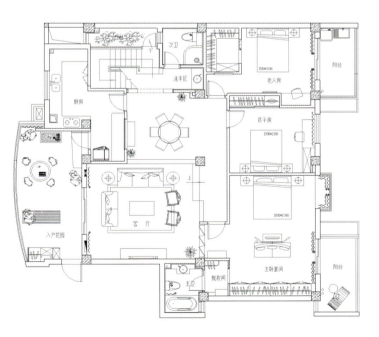

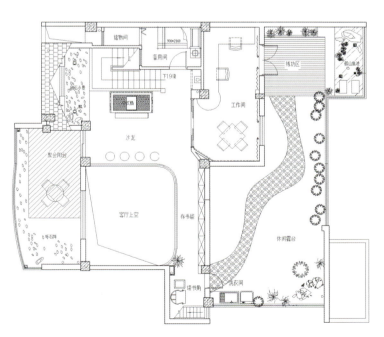

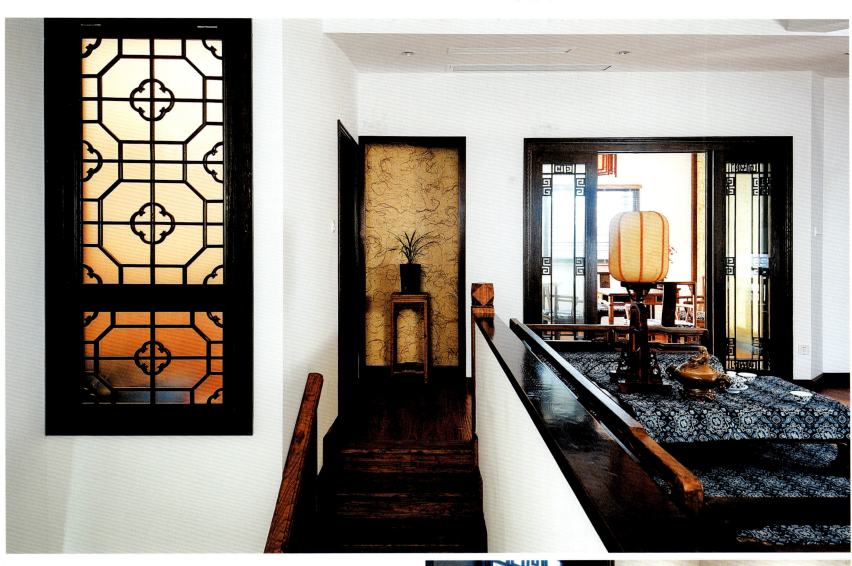

把一个钢筋水泥楼房变成散发着古韵的老宅，是设计师的设计初衷。设计师刻意使用中式建筑中的构件作为主要设计手段，希望能不留痕迹，不事雕琢，就营造出中式韵味。利用八角门、扇形窗、小姐窗，使各个空间互为借景，达到曲径通幽的效果。

Shandong Delicate Home—Shui Yue Yi Pin Show Flat

山东玲珑置业——水悦逸品样板间

The starting point of this project is to consider the functional practicability and full use of the space and the coordination of the overall. In this design, the traditional Chinese style is coupled with many modern elements, endowing the living space with more stylish feel. The collocation of colors is deep or light with the basic tone of light and elegant traditional Chinese color, partial embellishing generous mood with the gold color.

The living room attracts us with its warm and inviting visual enjoyment, and a whole of French window broadens our horizon.

The living room and dining room are open-style with flowing lines, showing a graceful and generous space from every corner. The smooth lines and minimalist styling are committed to simplifying the complex real life in the home design. The walls of balcony are decorated with wood, and the junctions of the walls and ground are decorated with pebbles, coupled with bamboo-decorated balcony handrail, highlighting a natural vitality of the space.

地点：山东　面积：120平方米　设计师：王锐　设计公司：沈阳一工室内设计事务所　主要材料：黑金花石材、茶色镜、手绘墙纸、金箔

  功能的实用、充分利用空间和整体的协调是设计师对本案设计的出发点。设计在传统的中式风格中注入许多现代的元素，使居室更有时尚感。室内色彩搭配有浓有淡，以中国传统的淡雅色彩为基调，在基础上局部点缀大气的金色等色彩。

  客厅空间给人以温馨怡人的视觉享受，整幅的落地窗拓宽了视野。客厅与餐厅的开放式布局，线条流畅，处处呈现落落大方的空间气势。流畅的线条、极简的造型都致力于将纷繁复杂的现实生活在家居设计中简化。阳台采用木质墙面，墙地交界处卵石处理与竹制阳台栏杆更突显了这一方空间自然的勃勃生机。

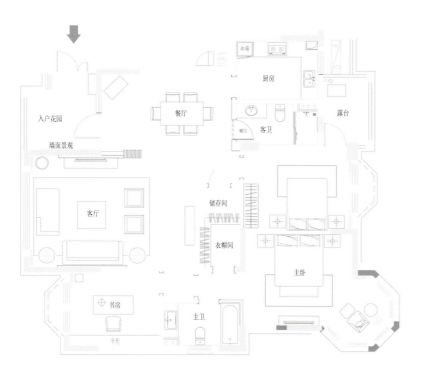

Conghua Moonlight Paradise—Southeast Asia

从化明月山溪——东南亚

In this project, a large amount of sago yellow stone, teak and other natural materials are introduced to create a comfortable and casual Southeast Asian style. The indoor lighting is very good, so that the space is spacious and bright. The rattan and teak furniture shows a fresh and natural atmosphere, and the wallpaper with flora patterns and the delicate patterns on the curtain coordinate organically to exude a winding and delicate sense of beauty. The square droplight in the dining room and the round one in the living room share similar material and complementary forms, playing the role of guide in their own space, highlighting the noble character of the space. A lot of fresh flowers and plants are added into the space, so in the space, we forget the noise of the city and feel as if dancing with nature.

地点：广州　面积：约170平方米　设计公司：广州万象整合装饰艺术设计有限公司　主要材料：西米黄石、柚木

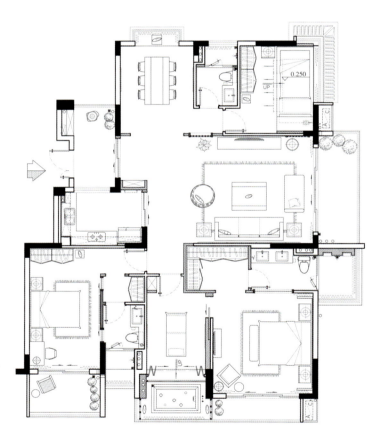

本案大量运用西米黄石、柚木等自然材质，营造出舒逸闲适的东南亚风格。室内采光性良好，使空间显得宽敞明亮。藤制和柚木家具铺展出清新自然的气息，花草纹样的壁纸和窗帘上的精致图案有机协调，绽放出蜿蜒细腻的美感。餐厅与客厅的水晶吊灯一方一圆，材质相似，形式互补，在各自的空间中均起到统领的作用，彰显着空间的高贵性格。室内还增加了许多鲜花植物的摆设，置身其中，让人忘却都市喧嚣，仿佛与自然亲密起舞。

New Olympic Town —Red Maple Garden

奥体新城 —— 丹枫园

This project is based on the Chinese-style aesthetics under the traditional culture, giving out a rich Oriental feeling. Added in a few tinges of colors and a touch of ornaments with Western-style elements, the interior space takes on a modern life style. As a consequence, in this single space, the blending of the two cultures is natural and harmonious, hospitable and unified.

The carved screen becomes "the time portal"; to one side of which, there is the comfortable and idle living room with New Urbanism, and to the other side of which, there is the antique-styled study that can bring you back to the ancient times when you could gaze upon the southern mountains leisurely. It allows the spirit of innovation to develop new elegance, which lets the lacquer, colors, wood and forms fully demonstrate the traditional Chinese strength and character, while on the other hand it employs fashionable colors and western-style decorations to enhance the rich modern feeling. Allow the straightforward log pattern to manifest the vivid appearance, and allow the soft fabrics to display gorgeous charm ... In reality, the seemingly paradoxical fashion and tradition, and the originally contrary Chinese and Western styles are the classic elements of mix-and-match aesthetics.

面积：130平方米   设计公司：南京传古装饰有限公司   主要材料：实木家具、实木地板、进口仿古砖、进口灯具

本案以传统文化下的中式审美作底，散发着浓郁的东方气息，加入几许时尚的色彩和一点西式元素的装饰，使室内有了现代的生活方式，在同一个空间里，两种文化的融合自然和谐，亲切统一。

雕花屏风变成了"时间之门"，一侧是舒适慵懒的新城市主义的客厅，一侧则是回到了古代悠然见南山意境中的古色古香的书房；让创新精神调理出新式优雅，一面让漆、彩、木、形尽现中国的传统风骨，一面用时尚色彩、西式装饰增添浓郁的现代色彩。让粗犷的原木纹理彰显灵动眉目，让温柔的布艺映现惊艳神韵……其实，看似矛盾的时尚与传统，原本对立的中式与西式，都是混搭在美学中的经典元素。

Corn Poppy

虞美人

This project applies Chinese-style hollow-carved pattern to the ceiling and walls separately, and the natural materials, such as Royal Botticino stone and Swietenia macrophylla King wood, have developed the simplicity and unsophisticatedness of the space into its fullest degree, which fully depicts the Chinese-style classical flavor. The harmonious blend of brown, coffee and yellow colors, along with the ancient-fashioned Chinese paintings and antiques, makes the whole space full of scholarly atmosphere. Alternated with modern elements of strong fashion styles, such as the plush sofa, the whole design reveals not only the Chinese-style feeling, but also the modern aesthetic perception.

地点：广州　面积：316平方米　设计公司：广州万象整合装饰艺术设计有限公司　主要材料：莎安娜米黄石、桃花芯木

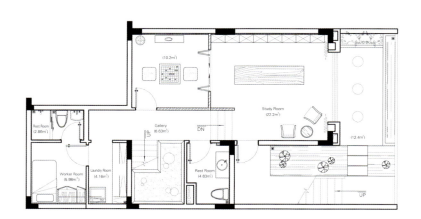

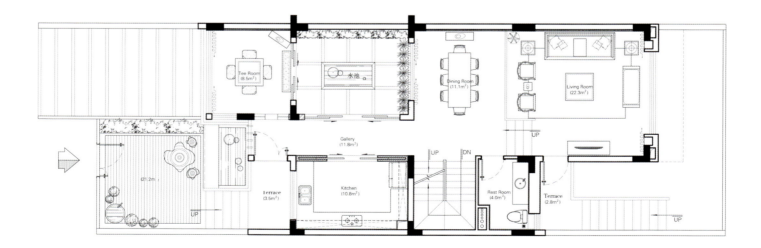

本案将中式镂空雕花图案分别运用于天花、墙壁等部位，莎安娜米黄石和桃花芯木等自然材质将空间的古朴素雅散发得淋漓尽致，尽显中式古典韵味。棕色、咖啡色和黄色三种色调和谐融合，再配上古色古香的国画和古董摆设，使整个空间充盈着书香气质。绒质沙发等颇具时尚感的现代元素穿插其中，使整体设计彰显中式风情之余，又不失现代美感。

Leisure Space

逸境

The designer chooses concise and soft colors, outlining a gentle home space with simple lines. At the entrance, the designer selects shutters and landscaping elaborately, creating an extended space for the residents to relax. In the living room, the warm beige light shoots down from the flat plain ceiling, so that the screen background wall exudes simple colors. The floor of the same color, fawn pure curtains, soft ground tiles, brown velvet sofa and other elements are collocated with each other harmoniously, so that the space is full of noble texture, highlighting the residents' high quality of life. The sofa background wall is decorated with simple beige wallpaper to add a tranquil atmosphere into the space, producing a sharp contrast to the translucent TV background wall.

地点：福州　面积：220平方米　设计师：蒋志烽　设计公司：福建国广一叶建筑装饰设计工程有限公司　方案审定：叶斌　主要材料：实木线条、艺术壁纸、艺术瓷砖、米黄石材、木格栅、艺术屏风

设计师选用了简明柔和的色彩，以简洁的线条勾勒出一个温存的家居空间。在入户玄关处，设计师精心挑选百叶帘与造景，为居住者创造一个舒展身心的外拓空间。在客厅中，平整的素色天花板上，温暖的米黄色灯光垂射而下，使屏风背景墙面散发出质朴的色泽。同色系的地板、驼色纯净窗帘、柔软的地砖以及咖啡色绒布沙发等元素的和谐搭配，使空间充满高贵质感，彰显着居住者高品质的生活。沙发背景墙用简单的米黄壁纸贴饰，为空间注入一份宁静的气息，与通透的电视背景墙形成了鲜明对比。

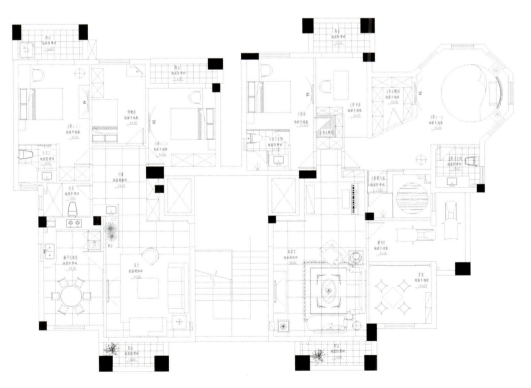

No.404, Building 2, Zhongxin Splendid Mansion

中信华府2幢404

The project emphasizes the experience of natural beauty with simple nature and modern technique, revealing a profound mood of elegance, adding a touch of culture, so we feel like encountering a classical taste of life.

In the whole space, the designer seeks the classical oriental flavor, at the same time, combined with simple techniques of modern design,he creates a relaxed and elegant atmosphere .

In the space, the designer chooses classic ancient furniture and accessories to demonstrate personality, and a lot of accessories with floral patterns of elegant plum blossom, orchid, bamboo and lotus are introduced into the interior environment, so we can capture the Chinese connotation easily at every corner. The ceramics and Chinese-style window frames serve as a contrast, and the elements in the space are independent, setting off an elegant realm of life.

地点：汕头　面积：197平方米　设计师：郭少雄　设计公司：汕头市铭庭装饰工程有限公司　主要材料：啡网石板、柚木饰板、日本墙纸、柚木实木地板

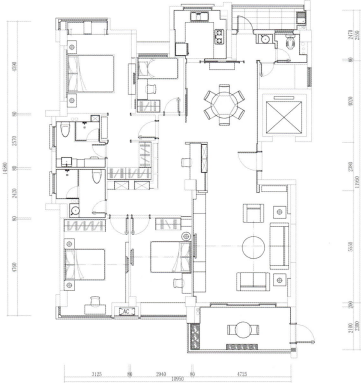

本案以古朴的本质、现代的手法着重体现对自然美的体验，优雅中透着深远的意境，增添了淡淡的文化气息，使人如同邂逅了一场古典的情趣生活。

整个空间在寻求东方古典韵味的同时，结合现代设计的简洁手法，营造轻松、雅致的氛围。在这个空间中，设计师以经典的古朴家具和饰品彰显个性，并将素雅的梅、兰、竹、莲等花卉图案的饰品大量地运用在室内环境中，让人于各个角落都能轻而易举地捕捉到中式的意蕴。在陶瓷、中式窗棂等元素的陪衬下，室内元素各自独立，烘托了高雅的生活境界。

No. 1602 Show Flat, Building No. 45, Hai Yue Garden—Modern Chinese Style

海悦花园45栋1602样板房——现代中式

The overall design of this project is the fresh light colors, whereas the furniture represents the rather sophisticated dark brown, which takes advantage of the partition measures of traditional Chinese-style space, so that the whole project is rich in arrangement and coherent in visual effects. This project has combined both qualities of sedateness and elegance. In terms of decoration, it employs the simplified and hard-edged straight lines, which represents an inventive designing concept, and takes on a Zen atmosphere. The whole project appeals a unique oriental charm, which takes the elegant and exquisite Chinese culture as the cultural deposits. For these reasons, this project is an integrated design which combines both the traditional culture and the fashionable elements, and vividly creates a refine and unpretentious lifestyle.

地点：顺德大良　面积：约120平方米　设计公司：广州万象整合装饰艺术设计有限公司

本案总体设计为清新的浅色调，家具则显现沉稳的深褐色，借鉴传统中式空间的划分手法，层次丰富，视觉连贯。本案融合了庄重与优雅双重气质，装饰采用简洁、硬朗的直线条，体现出独具匠心的设计理念，表达出一种禅的境界。整体空间散发出独特的东方韵味，以优雅细腻的中国文化为底蕴，是融合了传统文化和时尚元素的综合性设计，生动塑造了精炼而朴素的生活格调。

Forest in City, Shenzhen — Chinese Rhyme
深圳城市山林——中韵

"Chinese Favor" is the theme of this project, and the designer abandons complicated heaviness and absorbs modern conciseness, designing the whole living space with capable and experienced Chinese way. Around the theme, the designer starts from the collocation of basic materials, running the style throughout the furniture and ornaments.

The designer makes little modifications on the original building plan; his design focuses on the collocation of soft decoration. Through the introduction of screens, the tough wall becomes breathable, and the use of tea mirror endows the wall with more levels. Reflecting the concept of environmental protection, at the same time the "green" decoration is the base, following the designer's thoughts and feelings of life to show pieces of screens beautifully.

地点：深圳　面积：155平方米　设计师：刘升山　设计公司：百安居深圳南山分公司　主要材料：印尼茶镜、柚木地板、榆木屏风、热带雨林大理石

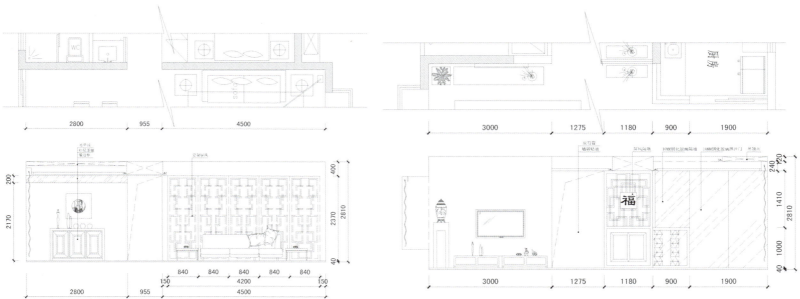

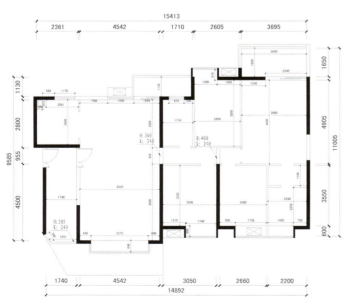

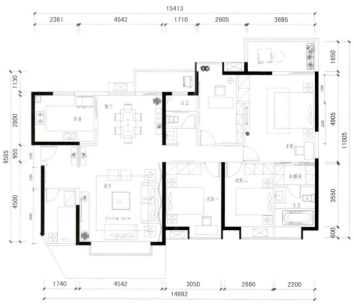

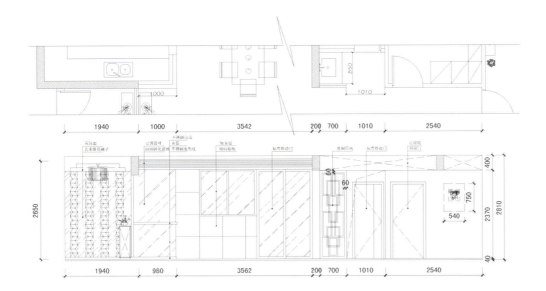

　　"中式情结"是本案的主题，设计摒弃繁复的沉重，吸收现代的简练，用中式的干练手法处理整个居室空间。设计师围绕主题从基础材料配置入手，将此风格贯穿家具饰品组合的始末。

　　设计师未对原有建筑平面做大幅度的修饰，而是将设计的重点放在了软装饰的搭配上，通过屏风的运用，使硬朗的墙体变得透气，而茶镜的使用也令墙体变得更有层次感。在体现环保概念的同时，以"绿色"装修为依托，跟随设计师对生活的感触，将一幅幅画面优美呈现。

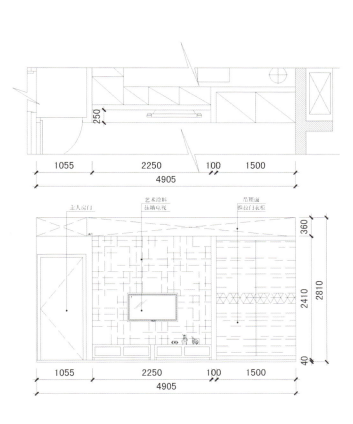

Chunjiang Garden

春江花城

This design shows its feature beyond our vision, clear and simple, with concern about the harmonious coexistence between nature and rationality. The extension of this design idea is not simply piling and flat placing, and the well-designed simple life creates a free experience for the owner.

The dark-colored shutters at the entrance effectively block the line of sight, meeting the functional need of air circulation, so the spaces are separated but not independent. The walls in public space are designed with plain-color stone to highlight the natural texture of stone. The living room is decorated with dark-colord furniture and carpet which are generous and simple, and the green plants are placed appropriately to fill the interior space with vitality. In the dining room, there are warm-colored lighting, orange walls, and pictures of summer flowers, exuding vitality and giving us a pleasant and warm dining experience.

地点：宁波　面积：150平方米　设计师：任朝峰、宋莹　设计公司：西点装饰　主要材料：大理石、墙纸、抛光砖

　　本案设计具有超越视觉的特点，清晰而简约，同时关注自然和理性间的和谐共处。此种设计思路的延展，不是简单的堆砌和平淡的摆放，精心设计的简约生活为居者营造一份自由的感受。

　　玄关暗色百叶隔栅有效地阻挡视线，满足空气流通的功能需要，使空间分而不隔。公共空间的墙面采用素色石材铺就，突出表现石材的本身质感。客厅空间采用深色家具和地毯，大气而简约。绿色植物的摆放恰到好处，令室内空间充满生命力。餐厅空间暖色灯光、橙色墙面、夏日花海的图片，散发着勃勃生机，给人愉快温馨的就餐体验。

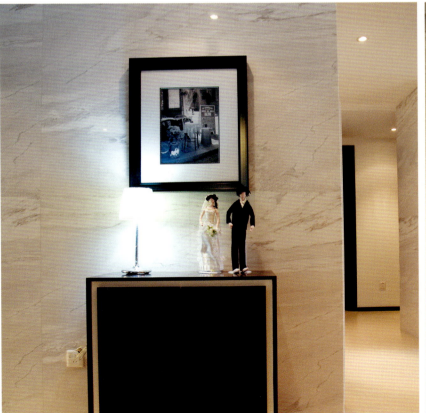

Starlit Court

星光华庭

The designer produces mental conversation with the owner through pure designing language, striving to create a pure land of soul for the owner in the noisy city. The designer breaks the traditional mindset, and pure white, wood color, and earth color are laid out naturally in the simple flowing space, back to nature. The stone-made dining table and simple droplight are quiet and elegant, discarding the distracting accessories and refusing the flashy modifications without substance. The glass sliding doors are sandblasted with geometric patterns, skillfully using the way of taking advantage of a scene in Suzhou Garden to enlarge the space, with separation but continuous, to create a transparent, smooth, and unique pure mood. The designer endows the limited pure space with infinite meaning, so that the owner will calm down in the loose and tight simple space, revealing a leisurely attitude towards life.

地点：金砂路　面积：131平方米　设计师：陈骏　设计公司：汕头市蓝鲸室内设计有限公司　主要材料：黑镜、喷白漆、微晶白石板、茶玻图案喷砂、面贴墙纸、木花银漆、茶玻图案喷砂　摄影师：邱小雄

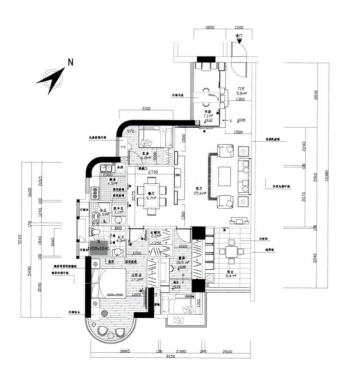

设计师以纯净的设计语言与居者心灵对话，力求在喧嚣都市里为居者营造一方心灵的净土。设计师打破传统思维定式，在流动的简约空间里，让素白、原木色、大地色系自然铺陈、返璞归真。摒弃喧宾夺主的配饰，拒绝华而不实的修饰，石质餐台及简约吊灯安静雅致。玻璃拉门上喷砂的几何图案，巧妙借鉴苏州园林隔而不断的借景手法开拓空间，营造了通透流畅、别具韵味的纯美意境。设计师将有限的纯净空间赋予了无限的意蕴，让居者在张弛有度的简约居室里沉静心灵，彰显着从容不迫的生活态度。

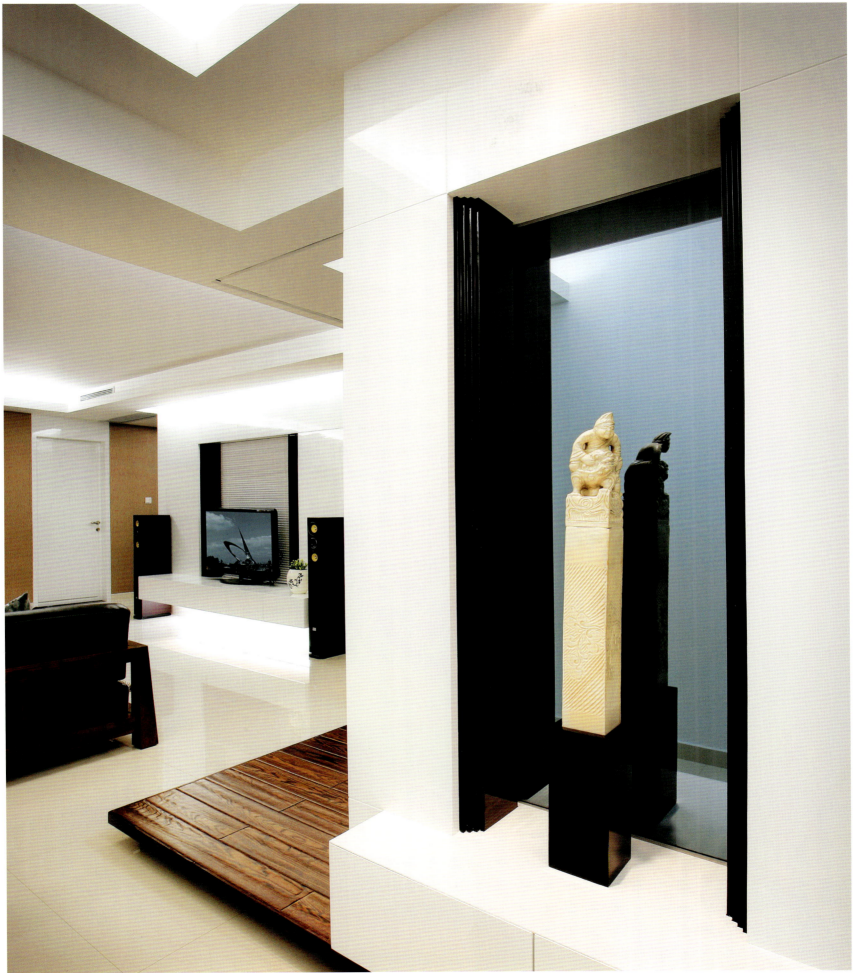

Southeast Asian Style

东南亚风格

Southeast Asian style represents a boundless enjoyment charm, in which the unsophisticated yet magnificent decorations appeal rather dignified and grand. The earth tone fills the whole space with warm aura. The wood material and parts of the wallpaper are complemented with marble, so that it reveals a modern, succinct, and bright vogue style. The whole residential design is not only modern in class, but also shows a special understanding of the Southeast Asian style. The rich arrangements in depths and the sharp contrast and harmony resulting from different materials provide a type of exploration and appeal for the development of modern people's residence. This project expresses an understanding of oriental feelings and an interpretation of the Southeast Asian leisure style.

设计师：黎剑　施工单位：贵阳中策装饰有限公司　主要材料：伊莎陶瓷砖、实木地板、隆森橱柜、汉斯格雅卫浴、雨林棕石材、艾尼得墙纸等

东南亚格调体现了无穷无尽的享受魅力，古朴而不失华丽温情的装饰显得高贵大方。大地色系让整个空间充满温暖的气息。木材与壁纸的局部配以理石，透露出现代、简洁、明快的时尚风格。整个房屋设计既现代又透出对东南亚风格的别样理解。丰富的层次感与材质强烈的对比与协调，为现代人居空间的发展做了一种探索和诉求。本案传达了对东方情愫的理解和对东南亚休闲风格的诠释。

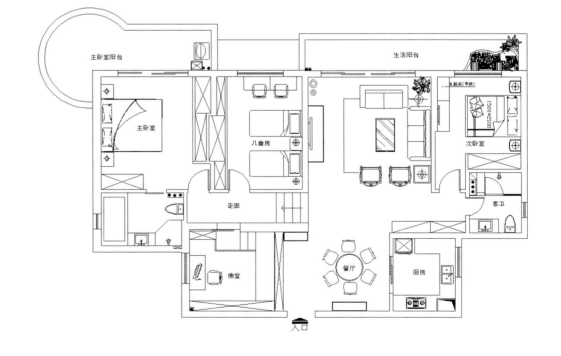

Love of Deep Fragrance

沉香之恋

The designer wants to unite taste sense and visual sense within one feeling, so he shows the calm and ancient rosewood with modern techniques, retaining the original flavor of wood, and he makes use of brilliant colors to create a modern atmosphere filled with assertive flavor.

Through visual treatment, the designer makes us smell a pleasant scent of wood, and the visual image has become the eternal taste reserved in our memory. The transition of warm and cool colors, the connection of heavy and light senses' every relationship exists in conflict, but the entire space is not out of balance. Different colors and materials coexist harmoniously and add radiance and beauty to each other.

地点：广东梅州　面积：165平方米　设计师：毛毳　设计公司：广东省梅州市澜庭设计工作室　主要材料：喷砂玻璃、艺术玻璃、白镜、花梨木、石膏板、乳胶漆、墙壁纸　摄影师：毛毳

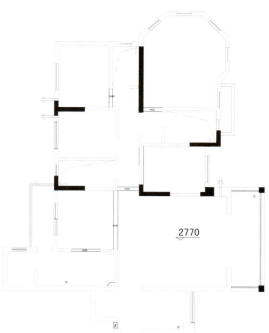

2770

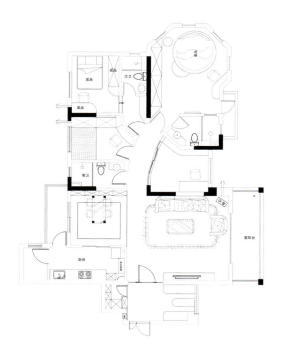

　　本案的设计师追求将味觉与视觉统一在一种感受之内，将花梨木的沉稳古色用现代手法去打造，保留那份木质的原味，又利用绚丽的色彩营造出充满张扬气息的现代氛围。

　　设计师通过视觉化的处理使人仿佛嗅到一份木质清香，使视觉形象成为了留存在记忆里的永恒味道。暖色与冷色间的过渡，重感与轻巧的衔接，每一个关系都在相互矛盾中存在着，但又不会使整个空间失衡。不同的色彩与材质在这一方空间中和谐共处，交相辉映。

Osmanthus City

桂花城

This project is defined as city talent's home, simple and stylish, with a little of national style. The design is combined with decorating style of traditional Chinese culture and modern design style, and the entire interior design strides over the national, modern and classic scopes. Through the designer's delicate design, the treated elements across many scopes are harmonious and bring out the best in each other. We can feel the designer's pursuit of tradition and steadiness from the interior details, and there are not many types of decorative elements, but they show their distinctive styles. The bedroom is designed on the theme of Southeast Asian style, cooperating with the Southeast Asian-style decorations in the living room and dining room. The plain-colored bedroom bedding makes the bedroom immerse in a quiet and peaceful atmosphere, with full display of natural flavor of life.

地点：杭州　面积：180平方米　设计师：李清河　主要材料：黑橡饰面板、抛光砖、橡木地板、柚木、丝绒窗帘

该方案被设计师定义为都市达人的居所，简约、时尚，又添加了些许的民族风。设计融合了中国传统文化和现代设计理念的装饰风格，室内整体设计跨越了民族、现代与古典的范畴，通过设计师的精心设计，使得这种跨界处理的元素和谐共处、相得益彰。在室内的小细节中更能感受到设计师对传统、沉稳的追求。各类装饰元素不需太多，但却个个风格鲜明。卧房以东南亚风格为主题，与会客、就餐区的东南亚饰品做一个贯穿。素色的床品更使卧室沉浸在一片安静、祥和的气氛中，一切尽显自然的生活气息。

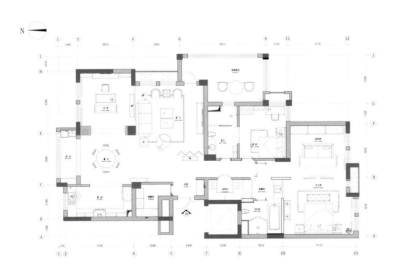

Charming East Asian Style

魅情东亚风

This project is a detached mansion of three storeys, and the Southeast Asian style home brings out Chinese elements in detail. The cylinder of columns in the hall is designed with round corners rather than the original square ones, and the structures echo perfectly and symmetrically. The living room is heightened partly with ramp-shaped ceiling on the top, breaking the square pattern and bringing vibrant atmosphere. The simple light-colored walls and heavy dark-colored wood are fully integrated in the design, showing the owner's high pursuit of living quality. The master bedroom and master washroom are completely open-type spaces, and the black marble organically integrates the floor and wall into one. The designer makes use of large-area window and mirror to expand our visual experience. The bamboo woven blinds, wooden figure of the Buddha, Thai teak woven panels, Thai-style fan lamps and other design elements are introduced to show an elegant and charming interior atmosphere of the space.

面积：320平方米　设计师：陈龙　设计公司：南京锦华装饰工程有限公司　主要材料：柚木、进口大理石、茶镜、仿古地板、乐家洁具、大金空调

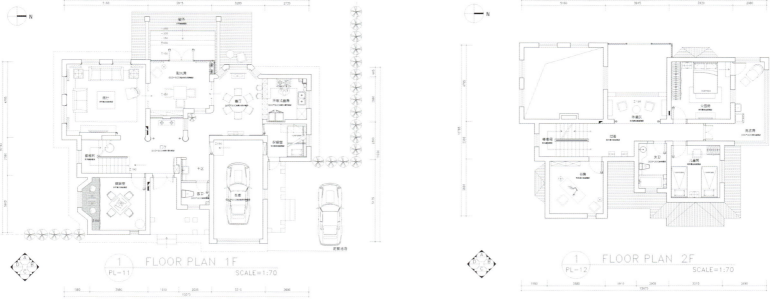

FLOOR PLAN 1F

SCALE=1:70

FLOOR PLAN 2F

SCALE=1:70

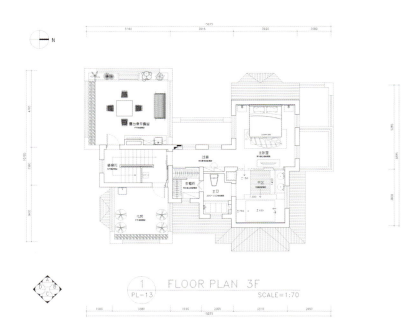

① FLOOR PLAN 3F
PL-13   SCALE=1:70

该案例是一个三层独栋豪宅，在这个充满东南亚风情的居所内，渗透着中式元素的细节。门厅立柱的柱面由原来的方角改为圆角，结构呼应、对称完美。客厅区域局部挑高，顶面设计成斜坡形吊顶，打破方正的格局，注入活跃的氛围。浅色墙面的简约与深色木料的厚重感在设计中完全融合，展示着居者对生活品质的高尚追求。主卧与主卫两者完全为开敞式空间，黑色大理石将地面与墙面两者有机地合二为一。利用大面积窗体及镜面等扩大了视觉感受。采用了竹制编织帘、木雕佛像、泰柚实木编织板、泰式风扇灯等设计元素，呈现出优雅迷人的室内氛围。

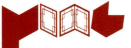

Taste New Chinese Style

品味新中

The designer introduces her own perception, accumulation and ideas to run this design, not only paying attention to the novelty of form, but also integrating her experiences of daily life into the design of the house, to make the living more convenient, more comfortable, and more enjoyable.

In the living room, the black leather sofa and black metallic droplight embrace with each other, showing a low-key and simple modern style. The hand-painted cabinet at the entrance, panes with lucky characters, blue and white porcelain mosaic, and red rustic fabric render the living space with rich vitality. The supporter with oblique lines in the living room breaks the prim visual impression of the space, bringing jumping visual experience. The original washroom is not divided into wet and dry parts, which is not convenient to use, so the designer changes the open direction of the washroom and outwards the wall to make a separate dry area, adding fun into the space.

地点：南京　面积：129平方米　设计师：于园　主要材料：楷模家具、玉石、艺术花格、马赛克、进口砖

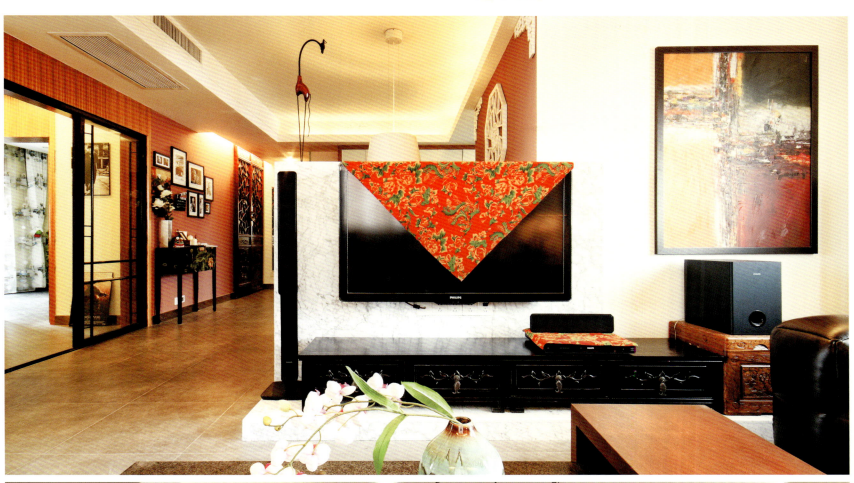

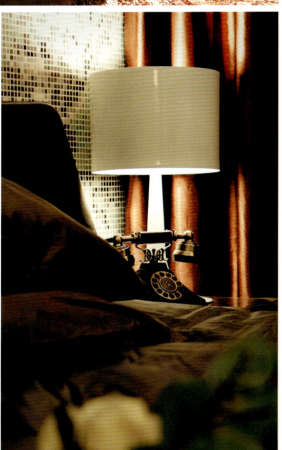

设计师利用自身感悟、积累、想法来处理此居室设计。不只是注重形式上的新颖，更将其日常生活的点滴心得融入到了居室的设计中，使居住更方便、更舒适、更惬意。

黑色皮质沙发与黑色金属质感的客厅吊灯相互辉映，体现出低调简约的现代风格。手绘玄关柜、福字窗格、青花瓷马赛克、红色乡村风格花布，把居室空间渲染得极富生命力。客厅中斜向线条的置物架打破了房子规矩的视觉印象，带来跳跃的视觉感受。原有卫生间干湿不明、不方便使用，设计师将卫生间门的方向改变，墙体外移，隔出独立的干区，更增加了空间的趣味性。

Peach Garden in Heyday, Chinese Style

盛世桃园中式

A lot of simple and tough straight lines are introduced into the space decoration of this design, and the use of straight-line decoration in space not only reflects modern citizens' requirement of pursuing simple life, but also caters to the Chinese design style of pursuing low-key and simple effects. The designers remove the wall of the study on the face of the door to make it an open-style study, and the conception of Chinese-style garden and modern materials are fully integrated, with brown wood battens inlaid transparent glass as partition, not only separating the study and the dining room, but also making the whole space more transparent and generous. In the living room, a kind of stone similar to light-color ink lines is introduced to a half-high TV wall, so that the landscape behind the wall can go into the living room. The furniture is brown mainly with simple lines.

地点：重庆　面积：106平方米　设计师：吴钒、吴扬武、梁瑞雪　主要材料：乳胶漆、墙纸、地砖、石材

本设计的空间装饰多采用简洁硬朗的直线条。直线条装饰在空间中的使用，不仅反映出现代人追求简单生活的居住要求，更迎合了中式追求内敛、质朴的设计风格。设计师打掉了正对户门的书房的墙，把书房改为开敞式。把中式园林的意境与现代的材质充分融合，以咖啡色木条作隔断，中间镶嵌透明玻璃，既把书房与饭厅做了分区，又使得整个房间显得通透大气。在客厅用类似于淡彩水墨纹路的石材做了一个半高的电视墙，让其后的风景融入到客厅中来。家具选用咖啡色系，以线条简练的为主。

Forest Town with Bright Sunshine
乡林阳明

The design of this project is like the ink character and white space in calligraphy, and the interior and exterior landscapes are designed with misplacing relations to set off the space style with the white space. Taking advantage of a scene is the technique, and the white space is the foreshadowing, adding the moving lines into the inside and outside spaces. The penetrating screens and grilles are used to define the current state of living space, and the opening and closing of the screens are used as the transitional hub of spaces. The space functions changes along with the opening and closing of the screens, creating an unique space concept, meanwhile meeting the demands for privacy and openness at different time point in the spaces. During the transition, a form of oriental space with imagination is created, like garden's twists when moving, with unique enjoyment. In the space, the designer highlights the artistic conception, texture, and romantic flexible beauty, blending classical and modern technical approaches to demonstrating the momentum and grace of the space.

地点：台北　面积：120平方米　设计师：杨焕生　设计公司：杨焕生建筑室内设计事务所　主要材料：格栅、秋香木皮洗白、烤漆铁件、大理石　摄影师：梁康为

　　本案的设计，如同书法里的墨字与留白，利用内景与外景作借位关系，用留白衬托空间风格。借景是手法，留白是伏笔，将动线书写于内外空间中。利用穿透屏风、格栅界定当下空间的生活状态，利用门扇的开合作为空间过渡的枢纽，开合间转变空间机能，产生特有的空间思维，同时满足空间中不同时间点上私密与开放的需求。转折间创造出一个具有想象空间的东方空间形式，移动间如园林般迂回，别富生趣。设计师在这个空间中强调了气韵与质感及浪漫的柔性美，融合古典与现代的技术手法，彰显居室的气势与雍容。

New Chinese Style

新中式

The designing strategy of Chinese style is introduced into the project, and some elements of modern style are integrated into the traditional style, so that the design does not adhere to the traditional Chinese style.

In the space, both the veneers and furniture are designed with large areas of solid wood, and the genuine color and texture make the interior atmosphere calmer. The pile of black bricks and the paving of stone create a contrast of style, but they harmoniously coexist in the denseness of the same atmosphere. The partition boards with Chinese patterns are used on the facade and ceiling, repeatedly setting off the theme of the space. In the washroom, the wash basin with blue-and-white pattern is bright, and it is the finishing touch. The design of the kitchen is also full of Chinese-style flavor, classic and subtle.

地点：南京　设计师：张磊　设计公司：南京动静空间设计工作室　主要材料：木地板、抛光砖、大理石、中式家具

　　本案采用中式风格的设计策略，传统中融合一些现代风格的元素，使设计不拘泥于传统中式的形式。

　　室内无论是贴面还是家具，均采用了大面积的实木，真实的色彩与质感使室内气氛更加沉稳。青砖的堆砌与石材的铺贴产生风格上的对比，却又在同一种气氛的氤氲之下和谐共处。中式纹样的隔扇形式在立面、吊顶等多处出现，反复烘托着室内的主题。青花纹样的面池色彩清朗，是空间中的点睛之笔。厨房的设计也充满了中式风情，古典而含蓄。

Rustic Oriental Rhyme

质朴东方韵

The designer chooses the traditional Chinese simple and rustic style, so that every corner of the space reveals an oriental flavor. A lot of partition boards with carved patterns and authentic Chinese-style furniture are introduced into the space to make the design elegant in Chinese style, not only full of flavor, but also full of the atmosphere of low-key fashion.

The design of the study is unique and interesting, and the small desk is placed in the corner, so we can feel the comfort of reading. The mahogany floor with Chinese flavor is full of festivity. The atmosphere is calm and interesting, and the seemingly tough lines are dotted with soft details. The washroom is designed with simple approach, refusing to imitate the style of retro decoration, and the classical elements are cleverly integrated into the modern decorating style.

地点：福州　面积：180平方米　设计师：蒋兴达　助理设计师：蒋兴通　主要材料：东鹏陶瓷、软包、大自然地板、L&D陶瓷

　　设计师采用了简洁质朴的传统中式风格，使室内处处呈现着东方的韵味。大量雕花隔扇的植入、地道中式家具的摆放，使设计在中式风格装修中透着一股典雅，不仅韵味十足，更具有一种内敛的时尚气息。

　　书房的设计别有趣味，小小的书桌呈现在角落里，更能让人感受读书的惬意，红木铺成的地板更具中国韵味，充满着喜气。沉稳中又富情趣，看似硬朗的线条中点缀着柔美的细节。卫生间则采用了简洁的手法进行处理，拒绝对复古装修风格的模仿，在现代的装修风格中巧妙地融入了古典元素。

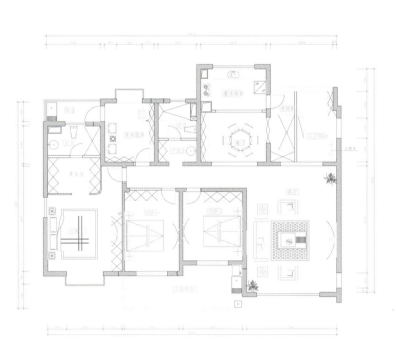

Charms of Tang and Song Dynasties
唐风宋韵

This duplex has large and tall rooms, but it does not have a symmetric format which is needed by the Chinese style and in the living room, the mobile line runs through the television wall. As a result, if it is not careful enough, it is prone to be disordered. The designer strives, in this grand architectural space with such a strong sense of modernity, with the partial embellishment of Chinese-style elements, to create this residence into an easy and practical home with a rich charm of Chinese culture.

Though the constructional area is not small, this duplex takes on a rather consistent plan format in up and down since it is built on the basis of the first and second floors of the apartment building. Taking into the practical needs of the family life, the designer places the living room on the first floor as the center of the whole residence. Part of the living room on the second floor, however, is connected with the bedroom of the owner; and part of the dining room on the second floor is taken as the tea room.

面积：300多平方米   设计师：冯克军

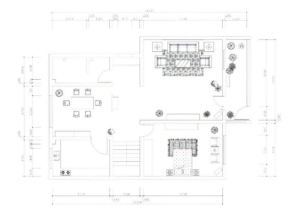

这套复式房空间较为高大，但又不是中式风格所需要的对称格局，再加上活动线贯穿客厅电视墙部分，所以稍有不慎，就会造成杂乱无章的局面。设计师力图在现代感极强的磅礴大气的建筑空间里，通过局部的中式元素的点缀，把该居室打造成一个方便、实用且具有浓郁中国文化韵味的家。

这套复式房虽然建筑面积不小，但是由于是在单元楼的一二层的基础上构建的，所以在整个结构上呈现的是上下一致的平面布局。考虑到实际家庭生活的需要，将一楼的客厅作为整套居室的中心。二楼的客厅部分则涉及业主卧室，而二楼的餐厅部分则考虑为茶室。

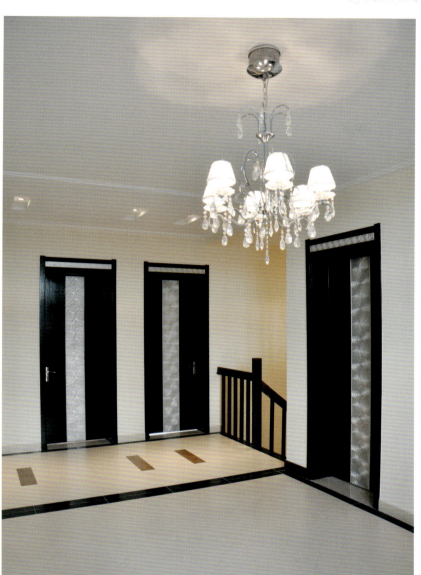

Zen-like Life

禅意生活

Oriental culture is the tranquil Zen that takes the self-adjustment and golden mean as its principal spirits. It advocates the simplified beauty in life, and will lead to a relaxation of both body and soul. The simplified style of the modern furniture, after experiencing all the prosperity, returns to its original nature, and gradually comes to the leading position in the designing trend of modern furniture, which manifests more of the spiritual needs of the modern people. In style, this residence is added with a personalized designing concept that is embedded with strong Zen culture, so that it reveals a harmonious, cool and natural Zen space. The elegant, graceful and plain Zen connotation, along with the chic and fashionable design, accomplishes a high standard of the interior space. Apart from that it also creates a spacious, simplified and aesthetic modern artistic space for the occupant.

地点：成都　面积：218平方米　设计师：郑军　设计公司：郑军设计事务所　主要材料：木饰面染色、墙纸、马赛克、仿古砖、大理石、实木复合地板

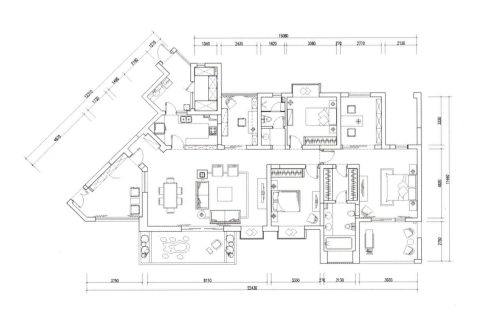

　　东方文化是以意欲自为调和、持中为其根本精神的悠悠禅意。它倡导生活中的简约之美，带来的是一种身心的解放。现代家居的简约风格在阅尽繁华后返璞归真，逐渐处于现代家装设计潮流的领先地位，它更多体现的是现代人精神上的需要。本套居室在风格中注入了带有个性化的浓重的禅意的设计理念，体现了和谐、冷静、自然的禅意空间。娴静、优雅、朴素的禅意内涵，前卫的时尚设计成就了室内设计的一种高境界，也为居住者创造出空灵、简朴、唯美的现代意向空间。

Residence Xiao, Tianmu International Village, Liyang

栗阳天目国际村肖宅

The designer spends a lot of energy on the detail design of facade, furniture and other aspects, not only taking the overall style, effect of practical expression, and the ease of use into account, but also considering the feasibility of construction. The living room and dining room communicate with each other, and the whole wall is designed with the same kind of real stone painting to extend the space continuously. The black painted glass is used on every layer, and the delicate feeling of vision makes the space for activities more spacious. The patterns of accessories are not traditional antique Chinese designs, but they pass out the flavor of traditional culture without exception. Coupled with fiddleback maple, beige painting walls, Chinese red and beige sofa, and cushions with colors, the rich colors with levels flow in the space, bring out the best in each other, and please us.

地点：常州　面积：266平方米　设计师：顾锋　设计公司：南京顾锋室内设计工作室　主要材料：地板、瓷片、真石漆　摄影师：顾锋

设计师在立面和家具等很多方面的细节设计上花费了大量的精力，不仅考虑到总体风格与实际表达的效果、使用上的便捷，更考虑到了施工的可行性。客厅与饭厅相通，整面墙面使用同种真石漆，让空间继续延伸。错层使用的黑色烤漆玻璃，视觉的微妙感受使活动的空间更显宽敞。配饰的图形虽然不是传统的、古色古香的中式图案，但无一例外地传递出传统文化的气息。搭配上白影木、米黄喷涂的墙面、中国红和米色的沙发、各具色彩的抱枕，丰富而有层次的色彩在空间中流动，相得益彰，使人赏心悦目。

2B Show Flat, Ou Peng Yu Jing Land Under Heaven

鸥鹏御景天下2B样板房

Simple, unsophisticated and sedate style is the type that is preferred by many occupants recent years. In China, most of the Chinese-style designs are the simple repetitions of each other; therefore, every design has to break through a lot of frameworks, otherwise they will not gain their occupants' favor. In this design, the designer boldly chooses modern style to serve as a foil for the Chinese theme, and applies the simplified and neat lines to separate the space into many a unit that has unique characteristics. The wood partition in the middle of the living room separates the dining room away from the living room, and the red Chinese-style furniture and decorations that can be seen everywhere are filled with pleasant atmosphere, so that the whole space is embellished warm and aesthetic.

地点：重庆　面积：160平方米　设计师：品辰团队　设计公司：重庆品辰设计　主要材料：地板、真丝手绘墙纸、木作着色、乳胶漆

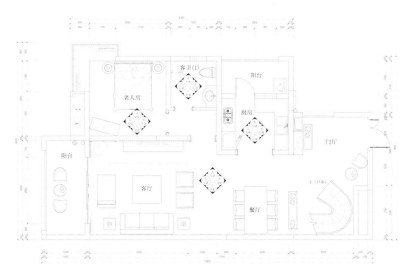

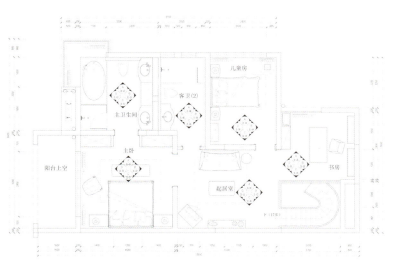

古朴稳重的风格是近年来很多业主喜欢的类型，在中国，中式风格的设计大多千篇一律，每一次设计都需要突破很多框架，才能获得业主的青睐。在本案中，设计师大胆使用现代风格衬托中式主题，简洁干练的线条将空间划分为多个独具特色的单元。客厅中央的木质隔断将餐厅和客厅分离出来，随处可见的红色中式家具和饰品洋溢着喜悦的气氛，将整个房间点缀得温暖、写意。

Rong Qiao Grand Residence
融侨华府

The design style of this project is the new Chinese-style classicism. The new Chinese-style decorative symbols have become the new designing elements that is appreciated the most around the world currently, which inherit the cultural deposits, historical aesthetics and artistic flavor of the classical oriental culture. The designer has simplified the complicated residential decorations into more succinct and elegant ones. He covers the straight lines with gentle and graceful soft decorations, and brings the classical aesthetics into the grand and practical modern designing techniques, so that the residence is more vivid, where the classical aesthetics penetrates the time and stays lively and vigorously by our side.

地点：武汉　设计师：朱勇　设计单位：清风组空间设计机构　主要材料：菠萝格长地板、定制花梨明清家具、西班牙云石灯、进口大理石、日本壁纸、金箔、欧洲布艺

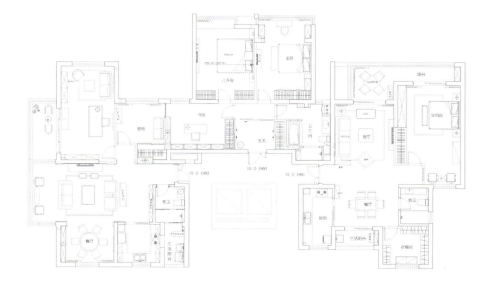

本案设计风格为新中式古典主义。新中式的装饰符号已成为当今国际最为推崇的新设计元素，它传承了古典东方的文化底蕴、历史美感及艺术气息。设计师将繁复的家居装饰凝练得更为简洁、精雅，为挺直的线条配上温婉、雅致的软性装饰，将古典美注入大方、实用的现代设计手法当中，使得家居更有灵性，让古典的美穿透岁月，在人们身边活色生香。

The Memory of Jiangnan

忆江南

The employment of black and white in traditional Chinese monochrome paintings has included everything in the world, and we are searching for the peaceful sheet of water at the bottom of our hearts in the balck and white. This project takes the Chinese traditional black and white as the basic line, which combines the bright moon, cyprinoids, and weeping willows in a scroll painting in the heart, whereat people can enjoy the glorious moon at the dining table, can view the fish lying on the couch, can take a casual stroll under the willows by the river, and can enjoy the enchanting river view in spring leaning against the beauty's support while reflecting upon a book. You will feel that there is no chaos of cities, and what you have is to let yourself calm down and enjoy life leisurely…

地点：福州　面积：170平方米　设计师：朱林海　设计公司：林海工作室　主要材料：通体砖、曲柳面板、玻璃、布面油画、壁纸　摄影师：朱林海

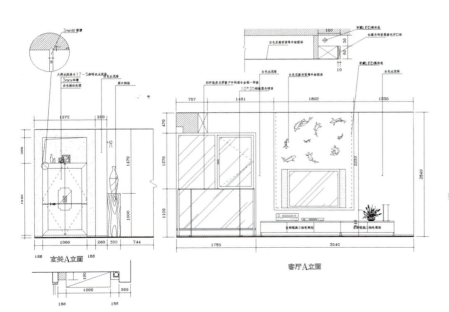

玄关A立面

客厅A立面

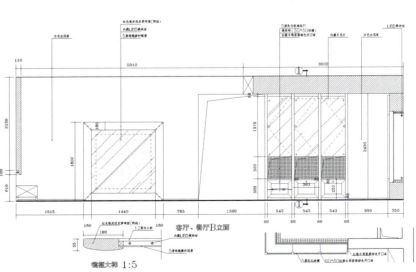

客厅、餐厅B立面

帽檐大样 1:5

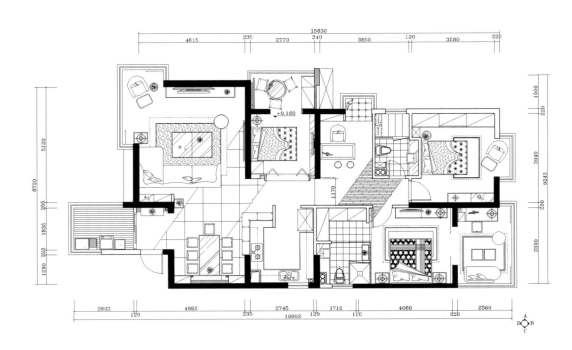

中国传统的水墨画中黑与白的运用把世间万物都概括起来了，在黑白中我们寻找内心深处那片平静的水面。本案以中国传统的黑白色为主线，把中国文化中的明月、鲤鱼、垂柳融合在一幅心中的画卷中，在那里，人们可以坐在餐桌边赏月，可以躺在沙发上观鱼，可以在河边垂柳下漫步，可以在美人靠上一边看着湖光春色一边拿着书慢慢咀嚼，心中没有城市的喧哗，有的是让自己静下心来细细地品味生活……

索 引

深圳城市山林——中韵

设计师：刘升山

设计公司：百安居深圳南山分公司

春江花城

设计师：任朝峰、宋莹

设计公司：西点装饰

星光华庭

设计师：陈骏

设计公司：汕头市蓝鲸室内设计有限公司

东南亚风格

设计师：黎剑

施工单位：贵阳中策装饰有限公司

沉香之恋

设计师：毛毳

设计公司：广东省梅州市澜庭设计工作室

桂花城

设计师：李清河

魅情东亚风

设计师：陈龙

设计公司：南京锦华装饰工程有限公司

品味新中

设计师：于园

盛世桃园中式

设计师：吴钒、吴扬武、梁瑞雪

乡林阳明

设计师：杨焕生

设计单位：杨焕生建筑室内设计事务所

新中式

设计师：张磊

设计公司：南京动静空间设计工作室

质朴东方韵

设计师：蒋兴达

助理设计师：蒋兴通

唐风宋韵

设计师：冯克军

禅意生活

设计师：郑军

设计公司：郑军设计事务所

栗阳天目国际村肖宅

设计师：顾锋

设计单位：南京顾锋室内设计工作室

鸥鹏御景天下2B样板房

设计师：品辰团队

设计公司：重庆品辰设计

融侨华府

设计师：朱勇

设计单位：清风组空间设计机构

忆江南

设计师：朱林海

设计公司：林海工作室